P9-EMC-075

ADVANCES IN FINANCIAL ECONOMICS

ADVANCES IN FINANCIAL ECONOMICS

Series Editors: Mark Hirschey, Kose John and Anil Makhija

HG
173
A383
2001
WEB

ADVANCES IN FINANCIAL ECONOMICS VOLUME 6

ADVANCES IN FINANCIAL ECONOMICS

EDITED BY

MARK HIRSCHEY
University of Kansas

KOSE JOHN
New York University

ANIL MAKHIJA
Ohio State University

2001

JAI
An Imprint of Elsevier Science

Amsterdam – London – New York – Oxford – Paris – Shannon – Tokyo

B3

ELSEVIER SCIENCE B.V.
Sara Burgerhartstraat 25
P.O. Box 211, 1000 AE Amsterdam, The Netherlands

© 2001 Elsevier Science B.V. All rights reserved.

This work is protected under copyright by Elsevier Science, and the following terms and conditions apply to its use:

Photocopying
Single photocopies of single chapters may be made for personal use as allowed by national copyright laws. Permission of the Publisher and payment of a fee is required for all other photocopying, including multiple or systematic copying, copying for advertising or promotional purposes, resale, and all forms of document delivery. Special rates are available for educational institutions that wish to make photocopies for non-profit educational classroom use.

Permissions may be sought directly from Elsevier Science Global Rights Department, PO Box 800, Oxford OX5 1DX, UK; phone: (+44) 1865 843830, fax: (+44) 1865 853333, e-mail: permissions@elsevier.co.uk. You may also contact Global Rights directly through Elsevier's home page (http://www.elsevier.nl), by selecting 'Obtaining Permissions'.

In the USA, users may clear permissions and make payments through the Copyright Clearance Center, Inc., 222 Rosewood Drive, Danvers, MA 01923, USA; phone: (+1) (978) 7508400, fax: (+1) (978) 7504744, and in the UK through the Copyright Licensing Agency Rapid Clearance Service (CLARCS), 90 Tottenham Court Road, London W1P 0LP, UK; phone: (+44) 207 631 5555; fax: (+44) 207 631 5500. Other countries may have a local reprographic rights agency for payments.

Derivative Works
Tables of contents may be reproduced for internal circulation, but permission of Elsevier Science is required for external resale or distribution of such material.
Permission of the Publisher is required for all other derivative works, including compilations and translations.

Electronic Storage or Usage
Permission of the Publisher is required to store or use electronically any material contained in this work, including any chapter or part of a chapter.

Except as outlined above, no part of this work may be reproduced, stored in a retrieval system or transmitted in any form or by any means, electronic, mechanical, photocopying, recording or otherwise, without prior written permission of the Publisher.
Address permissions requests to: Elsevier Science Global Rights Department, at the mail, fax and e-mail addresses noted above.

Notice
No responsibility is assumed by the Publisher for any injury and/or damage to persons or property as a matter of products liability, negligence or otherwise, or from any use or operation of any methods, products, instructions or ideas contained in the material herein. Because of rapid advances in the medical sciences, in particular, independent verification of diagnoses and drug dosages should be made.

First edition 2001

Library of Congress Cataloging in Publication Data
A catalog record from the Library of Congress has been applied for.

ISBN: 0-7623-0713-7

∞The paper used in this publication meets the requirements of ANSI/NISO Z39.48-1992 (Permanence of Paper).
Printed in The Netherlands.

CONTENTS

LIST OF CONTRIBUTORS

James S. Ang	Barnett Bank Professor of Finance, Florida State University, USA
Stewart L. Brown	Department of Finance, Florida State University, USA
Indudeep Chhachhi	Department of Accounting and Finance, Western Kentucky University, USA
Kathleen A. Farrell	Department of Finance, University of Nebraska-Lincoln, USA
Stephen P. Ferris	Department of Finance, University of Missouri-Columbia, USA
Kathleen P. Fuller	Department of Banking and Finance, University of Georgia, USA
Philip L. Hersch	Department of Economics, Wichita State University, USA
Mark Hirschey	School of Business, University of Kansas, USA
Tomas Jandik	Sam M. Walton College of Business Administration, University of Arkansas, USA

Elaine Jones — College of Business and Economics, Central Missouri State University, USA

Kari Jones — Department of Banking and Finance, University of Georgia, USA

Kenneth A. Kim — School of Business Administration, University of Wisconsin-Milwaukee, USA

Pattanaporn Kitsabunnarat — School of Business Administration, University of Wisconsin-Milwaukee, USA

Anil K. Makhija — Fisher College of Business, The Ohio State University, USA

Jeffry Netter — Department of Finance, University of Georgia, USA

Stuart Rosenstein — College of Business Administration, University of Colorado-Boulder, USA

Paul H. Rubin — Department of Economics Emory University, USA

Myron B. Slovin — Department of Finance, Lousiana State University, USA and HEC School of Management, Paris

Sridhar Sundaram — Emporia State University, USA

Marie E. Sushka — Department of Finance, Arizona State University and HEC School of Management, Paris

Edward R. Waller — Department of Finance, University of Houston-Clear Lake, USA

ON THE EXISTENCE OF SUB-STANDARD SECURITY MARKETS: THE IPOs OF BLINDER-ROBINSON

James S. Ang and Stewart L. Brown

ABSTRACT

In a standard security market, it is assumed that market participants are rational, capable of acquiring and utilizing information, fraudulent and opportunistic behaviors do not pay and are avoided, and transaction costs are relatively low. Furthermore, there is no group of participants who are consistent winners or losers; and collectively, no negative net present value projects are undertaken. We label a market as 'sub standard' if most of these conditions do not hold. Theoretically sub-standard markets are predicted to not exist at all, i.e. since even the most sophisticated institutional investors with the least transaction and information costs will not participate. This paper presents evidence of the existence of such 'sub-standard' security market over an extended period. Specifically, we present an in-depth analysis of twelve IPOs, underwritten by Blinder-Robinson. The IPOs were offered during a period when Blinder-Robinson was fighting unsuccessfully against NASD sanctions and SEC suspensions. The study is unique in several ways. It is the first detailed investigation of the market for the very small penny stock issues in which the per share offering prices

Advances in Financial Economics, Volume 6, pages 1–38.
Copyright © 2001 by Elsevier Science B.V.
All rights of reproduction in any form reserved.
ISBN: 0-7623-0713-7

of IPOs were at 5¢ or less and the gross IPO proceeds amount to only a few million dollars. We present evidence on the nature of the companies, the background of the executives and the board of directors and the business and financial characteristics of the IPOs. We also document the differences in the rates of return to various groups involved in the financing or management of these companies: the managers, other insiders, the underwriters and outside investors who invested before, at or after the IPOs. Finally, this study performs an original analysis of the microstructure of the penny stockbroker's operations by analyzing inside and outside bid ask quotes. In addition to broker's markups and markdowns, we have also calculated a measure of the market maker/brokerage house's incentive to its brokers to buy from or sell to their customers.

1. ON THE EXISTENCE OF SUB-STANDARD SECURITY MARKETS: THE IPOs OF BLINDER-ROBINSON

Security markets exist to facilitate trading of claims among strangers, to provide liquidity by minimizing transactions and search costs and to enable businesses to raise funds, etc. Underlying the foundations of the models of "standard security markets" are implicit assumptions and axiomatic propositions that are held to be generally true. For instance, market participants are expected to be rational and capable of making calculations by acquiring and utilizing information. In addition, it is assumed that market participants do not engage in transactions that are wealth reducing. To enable repeated transactions, some participants who have long horizons value records of previous performance and avoid fraudulent and opportunistic behaviors. Transactions occur in the market at relatively low cost because all participants perceive the benefits of trading as exceeding the costs. The expected as well as the realized returns from trading are non-negative. That is, there is no group of participants who are consistent winners or losers. There are no welfare reducing or negative net present value ventures successfully introduced into the market. And demand for securities does not go up with increases in supply simply because participants are unable to discriminate between true value and marketing efforts. When these conditions are not satisfied, the market for securities should theoretically not exist. The "standard" market is widely accepted as a fairly accurate characterization of existing security markets. Thus, the foundations for standard security markets may only be questioned if security markets violating these requirements are shown to exist.

In this paper we present empirical evidence on a security market that seems to have violated the premises of the standard security market. We label it as a sub-standard security market. The sub-standard market in question is the offering and trading of penny stock shares of IPOs underwritten by Blinder-Robinson.[1] These securities were initially priced at or under $0.05 (5 cents) per share and the average issue size was in the range of $3–$4 million.[2] Transactions costs, as measured by the outside (retail) bid/ask spread, are large. Information acquisition and analysis costs are high as these companies have little operating history and provide little data. There are also other implicit costs of trading as the underwriter is also the dominant market maker who sets the price and is often the only source of price quotations and is likely to be on the opposite side of each trade.

Institutional investors, the market participants who are the least handicapped trading in this market due to their lower transaction costs, strong bargaining position vis-à-vis the market maker, lower information acquisition costs and processing ability, do not trade in this market. Thus, in theory this market should not exist. (If the lowest cost participant does not participate, nobody should). And yet, it existed as these IPOs shares were sold to small investors, who had to pay high transaction costs, and made decisions with very little information, and thus, were predicted not to participate in the first place. Furthermore, we present evidence to show:

1. Welfare reducing investments were funded i.e. negative NPV IPOs are offered successfully, on a regular basis.
2. Even though the financial characteristics of the IPOs were of extraordinary high risk, the experiences of various groups were quite different. Certain groups of investors in the sub-standard market consistently earned positive returns. Some earned several times the amount of the original investment e.g. corporate insiders, underwriters and first day IPO flippers. On the other hand, other groups of investors consistently lost money i.e. outsiders who purchased shares prior to IPO, at one day after IPO and at post IPO offerings suffered heavy, if not complete, loss.
3. Marketing efforts on the supply side appeared to affect the amount of a security demanded. Market-makers such as Blinder-Robinson would change the payoff incentive to brokers in order to increase transactions rather than change the price. Analysis of micro-structure reveals that the daily adjustments fell mainly on the inside (wholesale) bid/ask spread, which affected the compensation to brokers, while the outside (retail) bid/ask spread change little, or did not change at all from day to day. Thus, our analysis of market microstructure reveals some unique aspects of the pricing and commission structure of penny stockbrokers. Of particular interest is the way Blinder-Robinson used inside and outside bid ask spreads to direct brokers efforts. We find that there was a much greater incentive given to brokers to have customers buy than to have customers sell.

4. Vital informations that could result in an unfavorable valuation of the shares, or might even cause the offering to fail were available, but apparently not utilized. For instance, there was adverse information such as pre-offering attempts to beef up the financials, questionable use for the funds raised such as for executive compensation or unspecified or "blind" purposes. Other relevant information that were apparently ignored include a great discrepancy between the estimated value of the issue obtained from the IPO companies own reported financial record and the offering price, and the adverse reputation of the underwriters where record of its serious legal difficulties with the security regulators, SEC and NASDAQ, that were also disclosed.
5. Outside investors also failed to take into account of the very high transaction costs implicit in the outside bid/ask spread. They are of such magnitude that, for the majority of securities, the prospect of breaking even when purchasing at the post offer first day price and after is practically non-existent.

2. DATA

We were able to obtain complete information of twelve IPOs brought to market by Blinder-Robinson (BR hereafter) in the period from late 1985 through late 1988. Information on the companies and financial statements prior to the IPO was obtained from S-18 and other disclosure statements filed with the SEC. Post-IPO financial information was obtained from SEC filings including 8K and 10K reports. Price information was obtained from broker quote sheets, pink sheets, the National Stock Summary and related information (10K, Moody's Industrial Manual). All information used in this paper is obtained from public records.

3. RESULTS AND ANALYSIS

3.1 The Characteristics of the Issuing Companies and the IPOs

Appendix 2 profiles the companies in the sample. The brief history traces the founding of the company, the nature of their business, the IPO and important subsequent events. In general the companies had been recently incorporated, had positive net worth but were not profitable. They represented a wide range of types of business. Three firms were in the food business (two restaurant chains and one distributing company). There was a toy company, a tool company, a finance

Table 1. Summary Issue Information for the Blinder-Robinson's IPOs.

The adjusted number of shares offered includes the original number of shares offered as reported in the offering prospectuses, the shares from the exercise of the underwriter's (Blinder Robinson) overallotment option, cheap stocks and warrants offered to the underwriters, and shares to be offered by the existing shareholders. Seafood's amended registration indicated that the major shareholder/president of the company was registering an additional 63,420,000 shares. Tele-Art is the only IPO with unit offering. Each unit is consisting of 1 share common stock, 1/5 share 'A' warrants at 100% premium over offer price, and 1/5 share of 'B' warrant at 200% premium over offer price.

Issuers	Date of Issue	Offer Price Per Share	Original No. of Shares Offered	Adjusted No. of Shares Offered	Gross $ of Proceeds Raised	Pre Offer Book Equity	% Share Retained by Pre-Offer Shareholders	Times Offer Price Per Share to Pre-Offer Book Value
1. Allertec	4/4/86	1¢	120,000,000	142,000,000	$1,420,000	$56,592	61.8%	14.7×
2. Amereco Environment	4/3/87	5¢	60,000,000	70,950,000	$3,547,500	$581,935	60.2%	9.3×
3. HDL Communications	3/3/87	2¢	112,500,000	132,187,500	$2,643,750	$230,624	70.0%	14.6×
4. Pasta Via	12/6/85	1¢	150,000,000	187,500,000	$1,875,000	$86,219	57.1%	13.4×
5. Paul's Place	11/12/86	1¢	170,000,000	203,150,000	$2,031,500	$36,867	65.3%	103.5×
6. Seafood Inc.	7/6/87	4¢	90,000,000	108,000,000	$4,319,200	$533,161	54.2%	13.7×
7. Tekna-Tool	12/7/87	5¢	100,000,000	118,000,000	$4,720,000	$705,574	61.3%	10.5×
8. Tele-Art	9/19/86	5¢	100,000,000	125,000,000	$5,750,000	$2,203,000	57.0%	4.77×
9. Thermacor	1/17/86	2¢	75,000,000	91,250,000	$1,825,000	$88,173	55.2%	25.5×
10. Trudy	7/14/87	4¢	100,000,000	118,065,000	$4,722,600	$1,350,368	63.2%	3.5×
11. Western Acceptance Corporation	9/17/87	4¢	62,500,000	70,746,480	$2,829,860	$761,520	50.5%	11.7×
12. Worldwide Bingo	8/6/86	1¢	150,000,000	179,625,000	$1,796,250	$153,375	55.3%	14.7×

company, a vanity press company, a specialty laboratory, an electronics manufacturer, a company that produced broadcast promotion packages and finally, a company that was developing a line of thermal reduction devices.

By 1993, all of the companies in the sample, with one exception, had either filed for bankruptcy, ceased operation and become inactive or had experienced mergers/buyouts which resulted in the original companies ceasing to exist, or even in the same line of business. The one exception is Tele-Art.

Table 1 lists the companies and presents summary information concerning the IPO issues. The issues were brought to market with offering periods between December 1985 and December 1987. Offer prices ranged from one to five cents per share with four issues at one cent, two issues at two cents, three issues at four cents and three issues at five cents.

The average number of shares in each issue was in excess of 100 million. Interestingly, this number is as many shares, if not more, than the typical security listed on the New York Stock Exchange. On average there were 10–20% more share actually offered than reported in the original offering prospectus. This extra supply of shares offered came about because of BR's overallotment (or greenshoe) option, and arrangements to allow BR to buy stock at deep discounts, i.e. cheap stock and warrants, and shares to be offered by the existing shareholders.

On average each IPO had gross proceeds of about $3.1 million with a range of $1.4 to $5.8 million. Thus, our sample is smaller than typically found in studies of small IPO, e.g. Guenther and Willenborg (1999). Interestingly, all of the companies show positive pre-offering book equity, which averages about $0.6 million. Thus, on average each offering raised between five and six times its original book value. (a later table will show that a significant portion of the pre-offer equity is due to pre-offering window dressing). In spite of this, the original owners retained the majority (60% on average) of shares and voting rights in the companies. The offering price to the purchasers of the IPO averaged twenty time the pre offer book value per share. The range was from 3.5 times (Trudy) to 103.5 times (Pauls Place). In general Table 1 shows that outside investors paid a rather high premium as measured by the high average offer price to pre-offering book value. Such high valuations associated with unproven small firms could only be warranted if future expected cash flows could justify them. This issue is examined in Table 5.

Table 2 profiles the issuing firms in the period before the IPO. The typical (median) firm had been in business only about six months. Firms were generally small with a median number of employees of ten people and were dominated by inside directors.

The characteristics and nature of the principals of the firms are often relied upon to highlight the legitimate nature of business enterprise of sample firms.

Table 2. Profile of Issuing Firms Before IPOs.

The source of information is the Form S-18, Registration Statement under the Securities Act of 1933. Since there are twelve firms in the sample, the median is calculated as the average of the sixth and seventh ranked firms. The number of employees is calculated on basis of full time equivalent where part time employment is counted as half full time.

		Median	Minimum	Maximum
1.	Months in Business Prior to IPO	6.5	3	111
2.	Number of Employees	10	2.5	1,100
3.	Board of Directors			
	Total	5	2	7
	Outside	1	0	2
4.	Auditors (n=12)			
	Majors			6
	Others			6
5.	Operating officer's background (n = 27)			
	Had responsible position in a major corporation			13
	Had relevant experience			14
6.	Officers and director's who were (n = 57) graduate of elite private universities			14
7.	Firms with officers or directors who were related (n = 12)			6
8.	Firms with Board members associated with venture capital firms			6

Half had auditing firms from the major accounting firms and about half of the operating officers had held responsible position in major corporations. About 25% of the officers and directors were graduates of elite private universities, and half of the firms had board members associated with venture capital firms or promoters. However, half of the firms had officers or directors related by blood or marriage to each other.

The picture that emerges from Table 2 is somewhat surprising. The issuers appeared to have identifiable products, experienced managers, some with good background, use mostly large auditing firms, and like most IPOs were not profitable mainly because they were less established, i.e. younger. Given the legal problems of BR at the time and the general poor repute of penny stock issues and firms,[3] the firms and principles in the sample appeared to be of surprisingly high quality.

Finally, we examine in Table 3 the announced intended use of funds from IPOs. It is of interests that a significant percentage of the funds (36.1%) are

Table 3. Proposed Uses of Proceeds from IPOs. The Proceed is Net of Underwriter's Fees.

	Uses	No. of Firm	Mean	Min	Max
1.	Product and process development including fees, review, permit, etc.	5	22.7%	16.3%	34.6%
2.	Product promotion: sales, marketing, advertising	9	22.3%	1.4%	41%
3.	Fixed asset investments	7	33.7%	8.8%	65.3%
4.	Various fees to consultants	8	3.2%	1.1%	6.1%
5.	Administrative cost	5	23.2%	13.1%	34.4%
6.	Working capital inventory	9	33.5%	8.2%	51%
7.	Reduce debt	2	12.5%	2%	23%
8.	Blind e.g. unspecified acquisitions	6	36.1%	13.6%	58.8%

designated for unspecified uses such as acquisitions and some are to payoff debt to insiders (average 12.5%) etc. The impact of this coupled with the relatively lower dollar commitments by insiders would further reduce the quality of the IPO for prospective investors and increase the risk inherent in the investment.

3.2 Pre-Offering Window Dressing

One of the reasons for the positive pre-offer book value of the firms in the sample is that there were significant efforts to beef up the book equity of the firms prior to the IPO. Table 4 analyzes the source of pre-offer book equity in the year prior to the IPO. Most of the firms had cumulative deficits prior to the offering and would have had negative equity had it not been for pre-offering maneuvers.

Table 4. Beefing Up Stockholder's Equity and Other Pre-Offering Maneuvers: A Summary of Increases in the Equity Account in the Period 12 Months Prior to IPOs.

Cash contributions from private pre-IPO offering of Equity				Non-cash Increases as % of Book			
	Shares in Private Offer[1] / Shares in IPO	Private Offer Price / IPO Price	$ Received / Pre-Offer Book Equity	Equity: Exchange Debt for Stock	Other Transactions[2]	Total Increase[3] to Equity	Pre-Offer Stock Splits
Issuers							
1. Allertech	145%	25%	117.3%	—	11.4%	11.4%	900:1
2. Amereco	—	—	—	17.1%	3.5%	138%	8,480:1
3. HDL	22.2%	36%	76.5%	124.7%	—	201.2%	1,446:1
4. Pasta Via	77.7%	13%	121%	72.3%	—	150%	1,102:1
5. Paul's Place	35%	17%	271%	144%	95%	510%	2,803:1
6. Seafood Inc.	4.7%	66%	17.4%	73.3%	68.6%	159.3%	5,000,000:1
7. Tekna-Tool	—	—	—	—	—	—	185.7:1
8. Tele-Art	—	—	—	—	—	—	—
9. Thermacor	109%	26%	99%	—	—	—	99%
10. Trudy	—	—	—	27.7%	—	27.7%	1,500:1
11. Western Acceptance	18.3%	66.4%	116.3%	55.7%	—	172%	2,822:1
12. World Wide Bingo	27.8%	17%	30.2%	—	—	30.2%	6,269:1

Notes: 1. Shares in IPOs include only the number of shares stated in the registration statement. Overallotment options and cheap stock granted to the underwriters are not included.
2. Examples of other non cash transactions are: (1) Issuer sold stocks to an affiliated company owned by insiders in exchange for a note (loan) to pay, the effects on the balance sheet are an increase in asset (notes receivable) and in equity; (2) Acquire a bankrupt company in a non cash transaction by agreeing to assume some debt, possibly at discount. The difference between the book value of the acquired/bankrupt firm and the assumed debt was recorded as an increase in asset and equity; (3) Issuers sold stock to an outside venture capital firm for cash, whose amount was equal or less than the contract fee for future consulting service; and, (4) Issued stocks in exchange for services rendered.
3. Issuers with sizable cumulative deficits could show new equity contributions, cash and non-cash, to exceed reported book equity prior to offering that included these additions.
4. Issuers that did not attempt to beef up equity account had sizeable equity prior to offering; computed as % of net offer proceeds, these are: Tekna Tool = 17.5%, Tele Art = 56%, Amereco = 21.6%, and Trudy 37.5%.

The issuing firms used various ways to pad their book equity. These methods included selling cheap stocks, i.e. at deep discount, debt to stock conversion and some non-cash, perhaps, questionable maneuvers.

Without the window dressing to book equity the pre-offering financial situation would have appeared to be even weaker than reported. In particular, the book values reported would be substantially less than those in Table 1, and the IPO offer prices would command even higher premia. The window dressing could contribute to the success of the stock issue offering and increase the amount raised.

3.3 A First Look at Ex-Post Results and Some Comparisons

A way to look at the "reasonableness" in the pricing of these penny stock IPOs is by comparing (i) what IPO investors paid to (ii) estimates of the value of the firms based on cash flows before and after IPOs. Table 5 presents the results of this analysis.

The imputed value of the firms (line 1) based on the IPO offer price (gross dollars at offer divided by the fraction of the total shares offered) had a mean value of about \$8 million for all twelve firms. The range was from \$3.8 million to \$16.8 million. Thus, based on the average, the total imputed value of the twelve firms was about 100 million dollars.

Lines 2 and 3 of Table 5 look at the value based on pre-IPO cash flows and lines 4 and 5 look at post-IPO cash flows. Based on the average of the two or three years of cash flows prior to the IPO the average firm value was about half a million dollars (450,000) with a range of 0.7 million to 5.2 million. Based on the average, a very rough estimate of the value of the twelve firms in aggregate is about 5 million dollars or roughly 5% of the value imputed from the offer prices. The corresponding number for the value based on the cash flows in the year prior to the IPO is about \$10 million or 10% of the value imputed from the offer prices.

Negative cash flows from operations in the years after the IPO (lines 4 and 5 of Table 5) resulted in negative firms values based on cash flows. The average liquidation value of the firms in the sample (exclusive of Tele-Art for which there was no data) based on the last financial statement filed with the SEC. was about 0.35 million dollars. Thus, the aggregate liquidation value for the twelve firms was roughly 4 million or 4% of the value imputed from the offer price.

It is also of interest to look at the financial results before and after the IPO in somewhat more detail. Table 6 compares sales, expenses, income and cash for two years prior to the IPO, the IPO year and three years post-IPO. Generally, part of the funds raised were used for expansion and thus sales increased.

Table 5. Estimates of Values Based on Cash Flows of Blinder-Robinson's IPO Firms Before and After IPOs.

		Mean (000)	Min (000)	Max (000)
1.	Imputed value based on IPO offer price: Gross $ at offer/fraction of total shares offered	$8,269	$3,824	$16,850
2.	Average of last 2 or 3 years cash flows prior to offer discounted at 20%	$450	$(771)	$5,224
3.	Cash flows of year prior to IPO discounted at 20%	$785.7	$(743)	$7,640
4.	Cash flows from operations of year after IPO discounted at 20%	(3,690)	$(7,235)	$263
5.	Cash flows from operations[1] over 3 years after IPOs	(2,376)	(6,225)	(763)
6.	Liquidation value, net of debt, based on the last financial statement filed with SEC (10K. 8K.) or market value[1]	$347	0	$2,000

Note:
1. These figures do not include Tele-Art.

However, expenses, or overhead increased at even faster rate, indicating either window dressing in the pre IPO years to improve income or reduced spending and later increased spending of "other investors" funds after the IPOs, a manifestations of the typical agency problem. The cash infusion from IPOs, however, did not help the firms to operate profitably. The net effect is that the cash that was brought in via IPOs was dissipated rather quickly and eventually most of the firms in the sample failed.

The value of the IPO firms, calculated either from pre- or post-offering cash flows or based on offer prices to the window dressed adjusted book value all indicate serious overpricing of these securities. In the following section we present more direct evidence of the investor's return.

Table 6. A Comparison of Some Financial Variables Before and After IPOs.

The information is taken from 10-K, 8-K, and Moodys. the IPO year refers to the fiscal year that included the IPO date. Due to the relatively large size of Tele-Art and its unique nature (Hong Kong), the figures are reported with (first figures) as well as without Tele-Art (second figures). Paucity of information for three years prior to IPOs precluded figures similar to these to be included.

		Sales	Selling, General and Administrative Expenses	Income	Cash
A.	Year prior to IPOs				
	2 years	$2,614	$436	($471)	$66
		$1,245	$366	($87)	$71
	1 year	$3,084	$512	$25	$67
		$1,512	$397	($18)	$69
B.	IPO Year				
		$3,446	$781	($574)	$1,126
		$1,857	$612	($300)	$862
C.	Years Post IPO				
	1 year	$5,121	$1,602	($954)	$341
		$2,799	$1,090	($816)	$231
	2 years	$6,347	$1,443	($563)	$417
		$4,191	$796	($447)	$124
	3 years	$3,691	$703	($499)	$382

3.4 Experience of Investors Who Purchased IPOs

3.4.1 At IPO

Table 7 shows the first day pricing of the twelve IPOs in the sample. In order to understand Table 7 it is necessary to define some terms. The data in Table 7 were obtained from internal BR manager quote sheets on the day of issue. The inside prices (bid/ask) are wholesale prices. For instance, for World Wide Bingo, the inside bid on the first day of trading was 4.75 cents and the inside ask price was 6.5 cents. These are the prices at which BR supposedly would trade on the interdealer market. The outside prices are retail prices to BR's customers. For instance, on the first day of trading the outside prices for World Wide Bingo were 4 cents bid and 8 cents ask.[4]

In Table 7, the average offer price on the twelve issues was 2.92 cents and the average outside ask price on the first day of trading was 7.15 cents, a 245%

Table 7. First Day Pricing of Blinder-Robinson IPOs.

	Offer Price Cents/ Share (% of Offer Price)	Outside Bid Cents/ Share (% of Offer Price)	Inside Bid Cents/ Share (% of Offer Price)	Inside Ask Cents/ Share (% of Offer Price)	Outside Ask Cents/ Share	Price Appreciation to Break* Even	Days ** Profitable
Allertech	1.00	2.00	2.75	3.25	4.00	100%	0
		200%	275%	325%	400%		
Americo Environmental	5.00	8.75	10.50	12.00	13.50	54%	0
		175%	210%	240%	270%		
HDL	2.00	3.50	4.50	6.00	7.00	100%	0
		175%	225%	300%	350%		
Pasta Via	1.00	1.90	2.25	3.00	4.00	111%	88
		190%	225%	300%	400%		
Paul's Place	1.00	2.00	2.00	2.50	3.50	75%	24
		200%	200%	250%	350%		
Seafood	4.00	5.75	6.50	7.50	10.00	74%	0
		144%	163%	188%	250%		
Tekna Tools	5.00	4.25	4.50	5.00	6.00	41%	0
		85%	90%	100%	120%		
TeleArt	5.00	5.75	6.75	7.75	10.00	74%	273
		115%	135%	155%	200%		
Thermacor	2.00	2.00	2.35	2.85	4.00	100%	0
		100%	118%	143%	200%		
Trudy	4.00	5.75	6.75	7.25	9.25	61%	0
		144%	169%	181%	231%		
Western Acceptance	4.00	4.25	4.50	5.00	6.50	53%	0
		106%	113%	125%	163%		
World Wide Bingo	1.00	4.00	4.75	6.50	8.00	100%	0
		400%	475%	650%	800%		
Mean***	2.92	4.16	4.84	5.72	7.15	72%	
		143%	166%	196%	245%		

* The price appreciation to break even is the first day outside ask price divided by the first day outside bid price less one. The outside bid price would have to increase by this percentage amount for the first day purchaser at the outside ask price to break-even.
** The Days Profitable are the number of days subsequent to the IPO where the outside bid price was greater than or equal to the first day outside ask price.
*** Mean percentages are the percentage of the average numbers rather than the averages of the %age numbers in the table.

premium over the average offer price. Thus, as the market maker, BR quoted a very large pricing jump (or "underpricing") on the first day of trading.

Recall from Table 5 that the aggregate market value of the firms imputed from the offer price was about 100 million dollars, which was about twenty times of a rough estimate of the aggregate value based on cash flow. The imputed total value of all firms at the end of the first day of trading was thus about 245 million dollars, based on the outside ask price. This is roughly 49 times the aggregate value of the firms based on cash flows or 250 times of a year's average cash flows.

In the IPO literature, the first day price jump is generally attributed as underpricing due to underwriter's aversion to risk, the desire to maintain a satisfied clientele of buyers, political legal liability or information asymmetry. Lesser known, however is the incentive for market makers to keep the offer price higher to enable it to sell its unsold inventory and/or exercising the overallotment shares.[5] Furthermore, the dramatic apparent increase in the market price could enable brokers to earn large commissions from trading on the first day.[6] Investors who were able to purchase the IPOs at the offer price or at prices less than the outside ask price[7] on the first day also did well. On average, if an investor (or flipper) received an allotment and was able to buy at the offer price and sell at the subsequent outside bid on the first day of trading , the rate of return averaged 43% for one day. At the same time the dramatic initial price increase and the riskiness of the companies involved resulted in a very high probability that investors who purchased the stocks in secondary trading after the initial offering would lose money. Consider the outcomes for investors who purchased at the outside ask (retail) price on the first day of trading. On average such investors paid 7.15 cents per share for securities which had an average outside bid (retail) price on that day of 4.16 cents. This means that on average the outside bid price would have had to increase by about 72% in subsequent trading for investors to break even and recover the initial purchase price. The lowest price appreciation to break even was 41% for Tekna Tools and the highest was 111% for Pasta Via.[8]

This dramatically high hurdle rate for price appreciation was momentarily possible for only three of the twelve securities in the sample. (For two of the three the break-even price existed only briefly so that investors would have needed nearly perfect timing in order to get their original money back). Stated differently, nine of the twelve securities never had outside bid prices equal to or greater than the outside ask price on the initial day of trading i.e. there was no single day these first day investors could break-even over the life of the stocks. The last column of Table 7 shows that number of market days after the IPO where the outside bid was greater than the first day outside ask price. Pasta via and Paul's Place were potentially profitable for 88 and 24 market days respectively. Only one security, Tele-

Art exhibited potential profitability for any appreciable length of time. This occurred in spite if the fact the Tele-Art continued to report losing money for two years after the IPO.

These results are shown graphically in Figs 1a, b & c which show the first 500 days of outside bid prices for all twelve IPO's. Also shown is a line parallel to the horizontal axis, which is the first day outside ask price.[9] It is of interest to note that the one company which did reasonably well, Tele-Art, was the company with the largest pre-offering book equity (2.2 million) and one of the lowest multiples of offer price over pre-offer book value (4.77× from Table 1).[10]

3.4.2 Longer Term Returns

It is possible to examine the profits and losses of different classes of outside investors from a longer term perspective. The results of this analysis are presented in Table 8 where undiscounted total cash contributions (–) and cash receipts (+) by each type of investor are compared. There, outside investors are classified into three groups; pre-IPO investors who purchased the shares before the IPO as informed investors and the shares were not registered, IPO investors and post-IPO investors who invested in subsequent rounds of equity financing. Interestingly, eight of the twelve companies were able to float another round of equity offerings after the IPO.

Even pre-IPO outsiders who paid discount prices did not fare well. Outsiders in three of the twelve firms got about 50% of their money back and the rest lost all of their investment. Six of twelve IPO investors got back an average of roughly ten cents on the dollar, investor in five of the eight firms with secondary offerings lost all of their money. The remaining three got back about fifteen cents on the dollar. In aggregate, all outside investors in six of the twelve firms lost all of their money and investors in the other six firms received ten cents on the dollar.

Investor's experience, when discussed in conjunction with the price analysis, show that in all cases, for all firms, and for all outside investors who invested at various stages of IPOs (before, at, or after), suffered heavy losses. The only class of investors who could possibly have made money were those who bought at the offer price or those who were able to buy at less than the day end's outside ask price and sell back the first day. The overall implication is that these penny stock issues were grossly overpriced at offer and more so at the end of the first day of trading.

3.5 The Experience of Company Insiders

Table 9 presents the sum of undiscounted cash inflows and cash outflows to various groups of outside investors. Cash flows to insiders include cash received

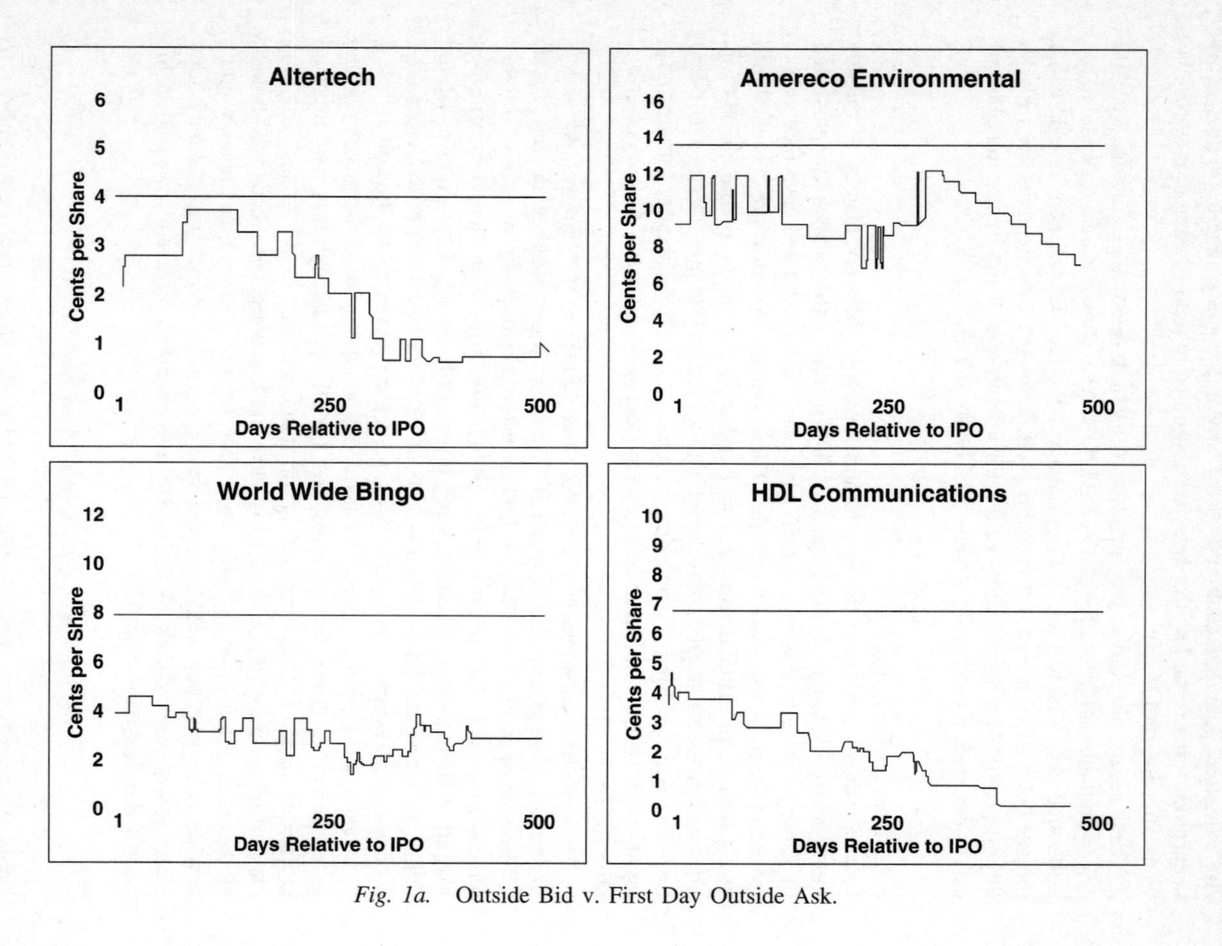

Fig. 1a. Outside Bid v. First Day Outside Ask.

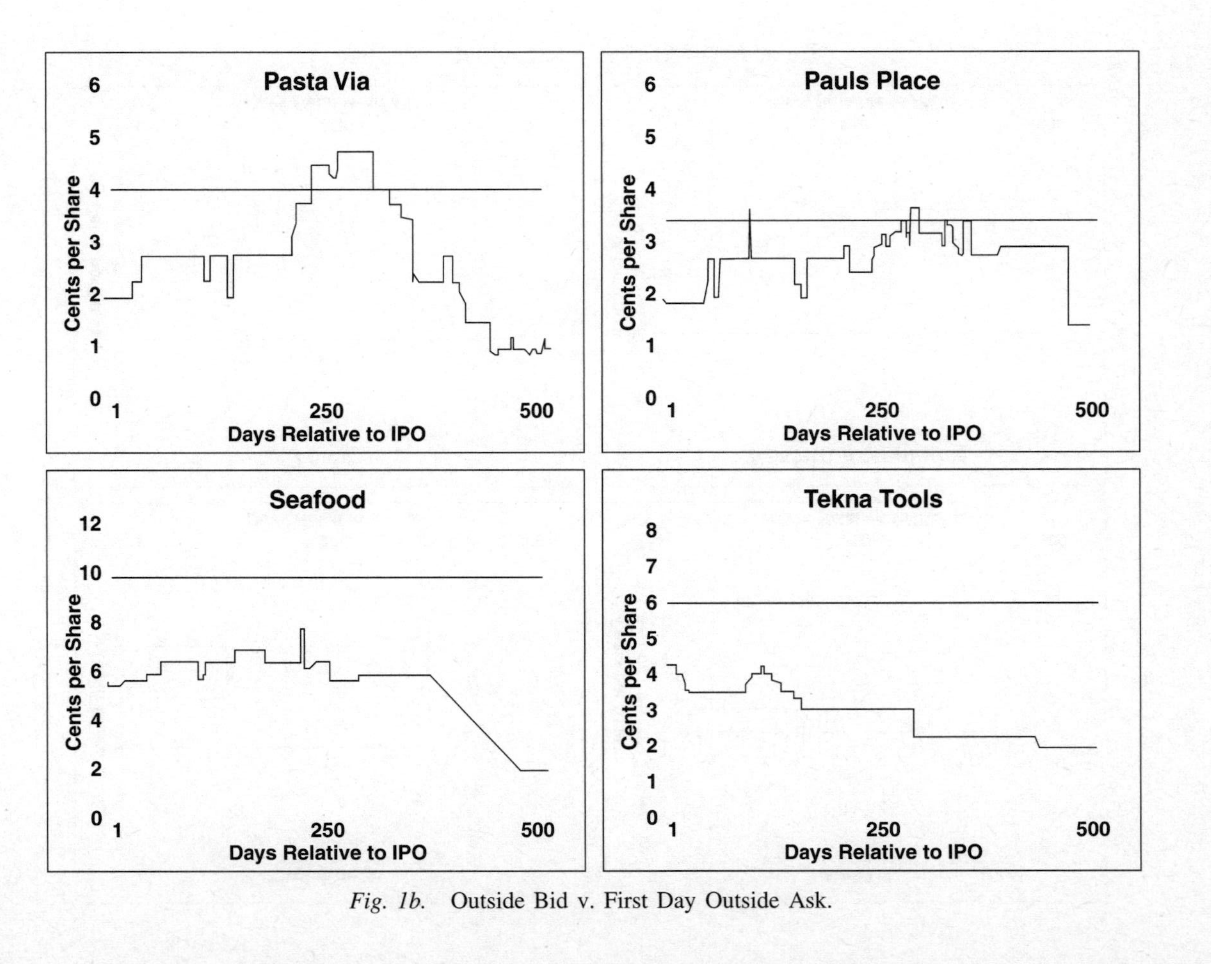

Fig. 1b. Outside Bid v. First Day Outside Ask.

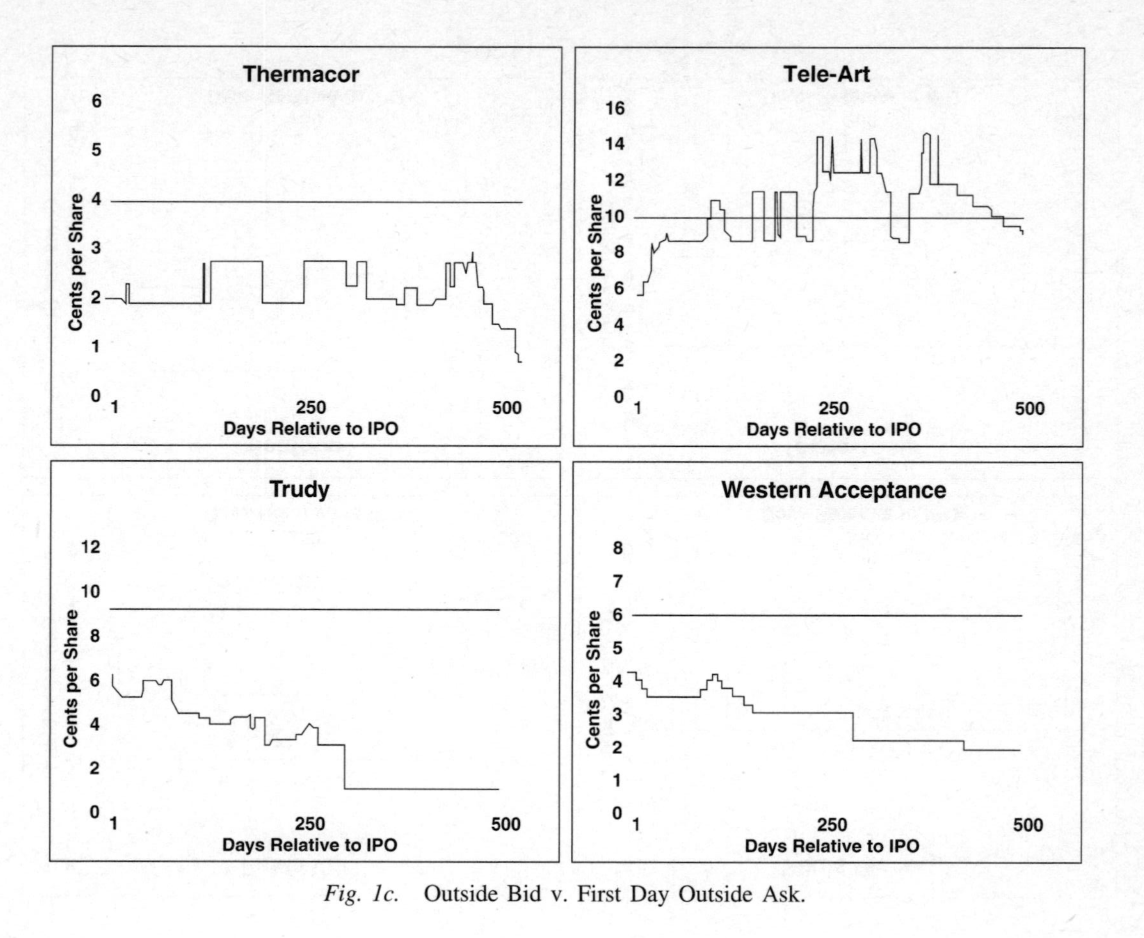

Fig. 1c. Outside Bid v. First Day Outside Ask.

Table 8. Gains and Losses to Classes of Outside Investors.

The table presents the sum of undiscounted cash inflows (+) and cash outflows (−) to various groups of outside investors. The investors are classified according to the stage of the firm when they committed their funds: before, at or after IPOs. The ratio (+/−) gives an undiscounted measure of dollar cash received per dollar invested.

Company	Outsider investors classified by stage of financing; purchase share at											
	All outside investors			Pre-IPO			IPO			Post-IPO		
	+	−	+/−	+	−	+/−	+	−	+/−	+	−	+/−
1. Allertech	$590,416	$2,821,614	0.12×	$107,468	$198,464	0.54×	$246,241	$1,420,000	0.17×	$236,000	$1,202,350	0.20×
2. Amereco	0	4,630,347	0	0	332,847	0	0	3,547,500	0	0	750,000	0
3. HDL	141,000	2,587,000	0.055×	-	-	-	141,009	2,587,000	0.055×	-	-	-
4. Pasta Via	0	2,803,255	0	0	208,200	0	0	1,725,000	0	0	870,055	0
5. Paul's Place	0	3,193,852	0	0		0	0	2,330,300	0		863,552	0
6. Seafood, Inc.	0	4,252,550	0	0	112,500	0	0	4,400,000	0	-	-	-
7. Tekna-Tool	1,636,800	5,390,000	0.3×	494,000	790,000	0.63×	1,142,799	4,600.000	0.25×	-	-	-
8. Tele-Art		9,416,000	-	-	-	-		5,750.000			4,766,000	
9. Thermacor	0	2,196,933	0	0	435,655	0	0	1,725,000	0	0	36,278	0
10. Trudy	3,121	4,722,600	0.00001	-	-	-	3,121	4,722,600	0.00001×	-	-	-
11. Western Acceptance	431,171	3,143,072	0.10×	43,200	314,072	0.14×	267,120	2,829,000	0.009×	784,800	4,100,000	0.14×
12. Worldwide Bingo	194,625	1,822,250	0.107×	15,000	26,000	0.58×	179,625	1,796,250	0.10×	75,000	750,000	0.10×

from sale of stocks, compensation (salary and bonus), consulting fees, severance pay, loans from the company which were forgiven or not paid and other identifiable benefits. Cash flows do not include perquisites such as the use of company cars or various forms of insurance. Cash outlays include cash paid for shares acquired and loans made to the company.

The picture that emerges on the returns to the insiders shows a dramatic contrast to the experience of outside investors. In practically all cases, insiders were able to at least recover their money and in most cases, returns were a large multiple of initial investment. Of particular interest is the group of insiders/professional investors (venture capital, blind pool promoters, etc.) who seem to have done very well for their relatively small exposure. Some were able to take over the firms when they were in distress later, e.g. Thermacor.

Blinder-Robinson also did very well. Table 10 shows the various sources of gains to BR up to the first day of IPO. Blinder-Robinson received compensation in various forms from underwriting these issues. There were underwriting discounts and commissions and non-accountable expense allowances. In eight of the twelve cases BR received rights to cheap stock worth a potential average value of $311,000 if sold on the first day of issue and in the remaining four cases BR received warrants worth an average value of $319,000 if exercised and sold on the first day of issue.

Subsequent to the issue BR received consulting contracts worth an average of $41,000 from each firm. The total compensation to BR is estimated to average about 0.75 million dollars per IPO, which averaged about 27% of the total average amount of each issue.

In addition to underwriting compensation, BR also profited handsomely from trading in the after market where it was the dominant broker. The gross commissions/mark-ups based on the first day bid ask spread for a single round trip average about 3.9 million dollars per IPO or a total of roughly 47 million dollars in aggregate. Thus, it is very likely that BR generated greater profits from trading in the after market than the IPO proceeds.

This explains why underwriters were willing to provide the underwriting service to the small penny stock issuers. The non-negative returns to insiders explains why firms were willing to sign up with BR in spite of its known very serious legal problems.

It may be argued whether sufficient adverse information was adequately disclosed to potential shareholders. Table 11 details the types of known risk as reported in the registration statements. The picture that emerges is that all of the IPO issues were very risky and this was disclosed in some detail in the registration statements. The small number of shareholders needed to subscribe to an IPO issue, and the large number of Blinder's brokers could provide an

Table 9. Gains and Losses to Classes of Insiders.

The table presents the sum of undiscounted cash inflows (+) and cash outflows (−) to various groups of inside investors. Cash inflows to the insider include cash received from sale of stocks, compensation (salary and bonus), consulting fees, severance pay, loan from company (forgiven or not paid), and other identifiable benefits, but do not include perks such as use of company car, health insurance, life insurance policy, and liability insurance for officers and directors. Cash outlays include cash paid for shares acquired, loans made to company that are still outstanding or converted to equity.

Company	All Officers and Directors or Identified as Promoters			Unaffiliated Officers & Directors			Officers and Directors Who are Affiliatedwith Venture Capital Firms, Business Consulting Firms, Blind Pools		
	+	−	+/−	+	−	+/−	+	−	+/−
1. Allertech	$1,379,623	$117,760	11.723	$1,239,271	$53,681	23.01×	$140,352	$64,079	2.19×
2. Amereco	760,972	101,970	7.47×						
3. HDL	837,000	416,456	2.01×						
4. PastaVia	832,280	74,620	11.15×	54,184	27,270	2.0×	778,096	47,420	16.41×
5. Paul's Place	1,169,000	1,170,553	1.0×	68,000	35,000	2.0×			
6. Seafood,Inc.	1,133,839	314,500	3.60×						
7. TeknaTool	2,350,000	332,000	7.08×						
8. Tele-Art	2,203,000	1,600,000	1.38×		600,000	2.67×			
9. Thermacor	1,662,700	192,000	8.63×	82,000	0	–	1,580,700	192,600	8.21×
10. Trudy	1,690,230	1,486,958	1.0×				2,075,000	29,200	12.8×
11. Western Acceptance	2,177,500	526,700	4.13×				102,500	497,500	0.21×
12. Worldwide Bingo	374,000	17,699	21.13×						

Table 10. Sources of Compensation to Blinder Robinson From IPO Underwriting.

I. At Date of Issue or Prior	Mean	Min	Max
a. Underwriter discounts/commissions as % of issue	10%	10%	10%
b. Nonaccountable expense allowance[1]	3%	3%	3%
c. Underwriter fee from exercise of overallotment option.	1.5%	1.5%	1.5%
d. Right to cheap stock or warrants			
1. Stocks (N=8)			
a. Price Blinder Robinson paid as a fraction of IPO price	1/216	1/500	1/100
b. Profit to Blinder Robinson if exercised and sold on first day	$311,125	$179,700	$569,288
2. Warrants (N=4)			
a. Premium of warrant exercise price over offer price	.20	.20	.20
b. Profit to Blinder Robinson if exercised and sold first day	$319,000	$120,000	$420,000
II. After Issue			
a. Investment banking consulting service contract[2]			
1. Dollar Amount	$41,417	$24,000	$48,000
2. Length in months	21.25	10	24
b. Other contractual terms			
1. Right to attend board meeting as observer			
2. Preferential rights to future financing			
III. Total compensation, fees, and profits to BR around time of Issue			
a. Dollars per IPO	$756,000	$392,000	$1,465,000
b. As a % of issue amount	27.2%	18%	47.7%
IV. Gains from market making as the dominant broker in after market			
a. Gross commission, based on first day bid/ask spread for only a single round trip cross trade.	$3,878,000	$1,617,000	$7,094,500
b. (a) as a % of issue amount[3]	146.7%	43.75%	400%
V. Gain from exercise of warrants granted in unit offering,[4]	$1,000,000	$1,000,000	$1,000,000

[1] The issuing firms typically paid Blinder-Robinson $25,000 in advance before the issue.
[2] Nine if the twelve IPOs had the standard $48,000 (12 month)
[3] In ten of the twelve issues the total fees and the gross commissions from a single round trip transaction to Blinder Robinson exceeded the gross issue amount to the issuer.
[4] Blinder Robinson was paid $600,000 to exercise 10,000,000 registered shares of Tele-Art that was worth a market value of $1,600,000 at the interbroker price. Tele-Art spent $189,000 to pay for Blinder Robinson's management tour to Hong Kong.

Table 11. Types of Risks in the Registration Statement.

I. Common risks, as warnings found in the registration statement

a. The IPO price has been arbitrarily determined by the company and the underwriter, and bears no direct relationship to the assets. Earnings, book value or any other objective standard of worth.

b. The securities are highly speculative, involve significant risk and immediate substantial dilution, and should not be purchased by any person who cannot afford the loss of his entire investment.

c. Discuss extensively (1-2 pages) NASD and SEC actions against Blinder-Robinson, and the possibility that it may nor be able to perform as underwriter.

d. To indemnify each other, underwriter/issues, against liabilities incurred by reason of mistatements or omissions to state material facts in connection with statements made in the registration statement and prospectuses.

e. Warn of poor liquidity of shares, and Blinder-Robinson's dominant role in market making.

II. Issuer Specific Risks	Number of Issues	Number of Instances
a. Civil lawsuits against issuer	5	10
b. Action from federal (IRS,Labor Dept. EPA) State agencies	4	6
c. Officer's background: Felony charges, license suspended (as broker), declared personal bankruptcy	4	5
d. Near term liquidity crisis: low cash, bank debt due, negative working capital, negative bank balance	10	10
e. Product/business weaknesses: no product at developmental stage, loss of major account potential liabilities, hidden liabilities from guarantee given to affiliate.	4	4
f. Loans to officers	5	7
g. Business dealings with officers and directors: purchases of goods, services and consulting but exclude legal services.	6	7

	Mean	Min	Max
h. Use of funds to increase management compensation: (N=11)announced expected increase	+254%	+29%	+970%

explanation why such overpriced and risky shares could be sold, i.e. a couple of thousand brokers looking for a few hundred or couple of thousand investors per issue via high pressure sales tactics. Thus, it may not be a question of adequate disclosure but rather it may have to do with the investors' lack of ability or willingness to do their own homework (versus relying on Blinder's brokers who may have an agency conflict) and the timeliness of the availability of this information.

4. MARKET MICROSTRUCTURE OF BLINDER-ROBINSON PENNY STOCKS

Figure 2 shows the average bid-ask prices for all twelve IPOs for approximately 3 years (750 market days) after the IPO date. Initially prices increased slightly and average outside ask price were up about 15% in the first 90 market days or about 4.5 months; recall there was a policy to encourage cross-trades during this period. After about a year, average outside ask prices were essentially flat as compared to the first day of trading. After 500 market days or two years, average outside ask prices were down about 35% and 65% after 750 market days or three years.[11] Average outside bid prices moved essentially parallel to outside ask prices.

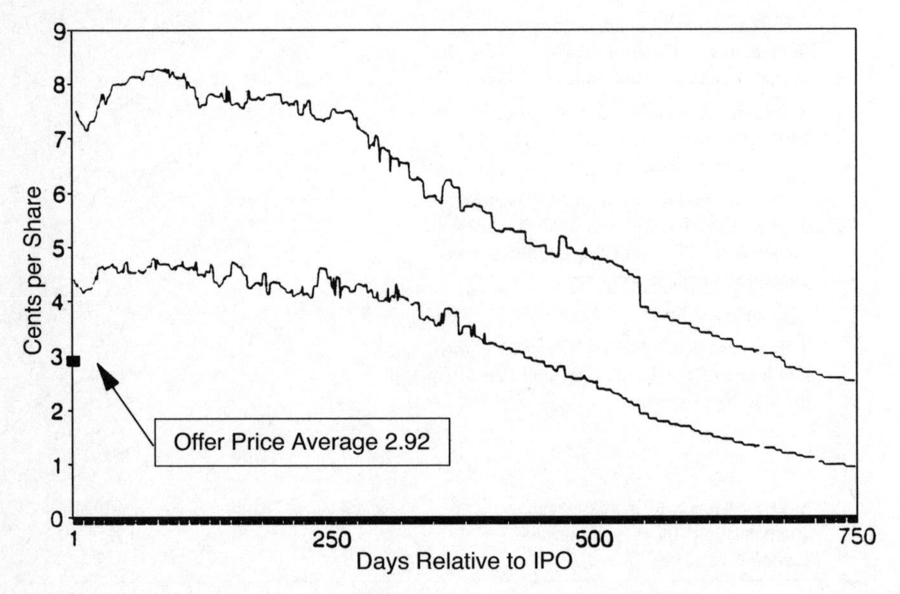

Fig. 2. Average Bid-Ask Prices Relative to IPO Date.

The above analysis, though interesting, is obviously flawed because, as noted above, investors had to sell at lower prices than they paid. Figure 3 shows the average spread between outside bid and ask prices for the twelve BR IPOs relative to the IPO date. On average, spreads were 3.31 cents per share during the first year and 2.4 cents per share during the second year. Although absolute spreads decreased from the first to the second year of trading, relative spreads were a fairly constant at 74 and 70%, respectively, of average outside bid prices, and at 43 and 41%, respectively, of average outside ask prices. While spreads decreased, so did prices so that percentage spreads remained relatively constant. Thus, on average, investors could only sell their purchases for about 60% of cost.

Blinder-Robinson's average mark-up and mark-down on the securities it traded may provide evidence to support the hypothesis that BR encouraged its brokers to sell securities but not to buy them back from customers.[12] If A_o and A_i are the outside and inside ask prices respectively and B_o and B_i are the outside and inside bid prices respectively, the percentage mark-up may be defined as $(A_o - A_i)/(A_i + B_i)/2$ and the percentage mark-down may be defined as $(B_i - B_o)/(A_i + B_i)/2$. The denominator of each fraction is the average of the inside bid-ask prices to represent the contemporaneous prevailing market price

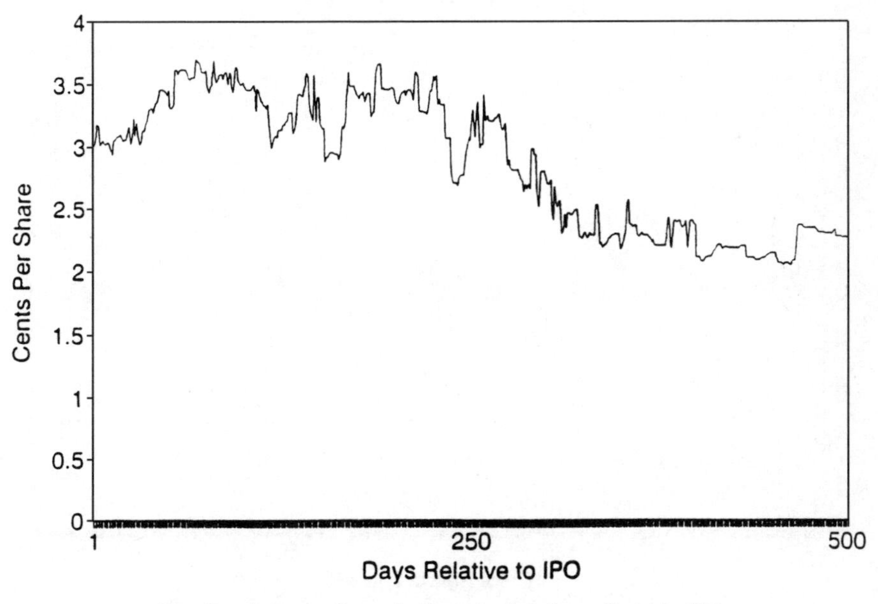

Fig. 3. Average Spreads Outside Ask Less Outside Bid.

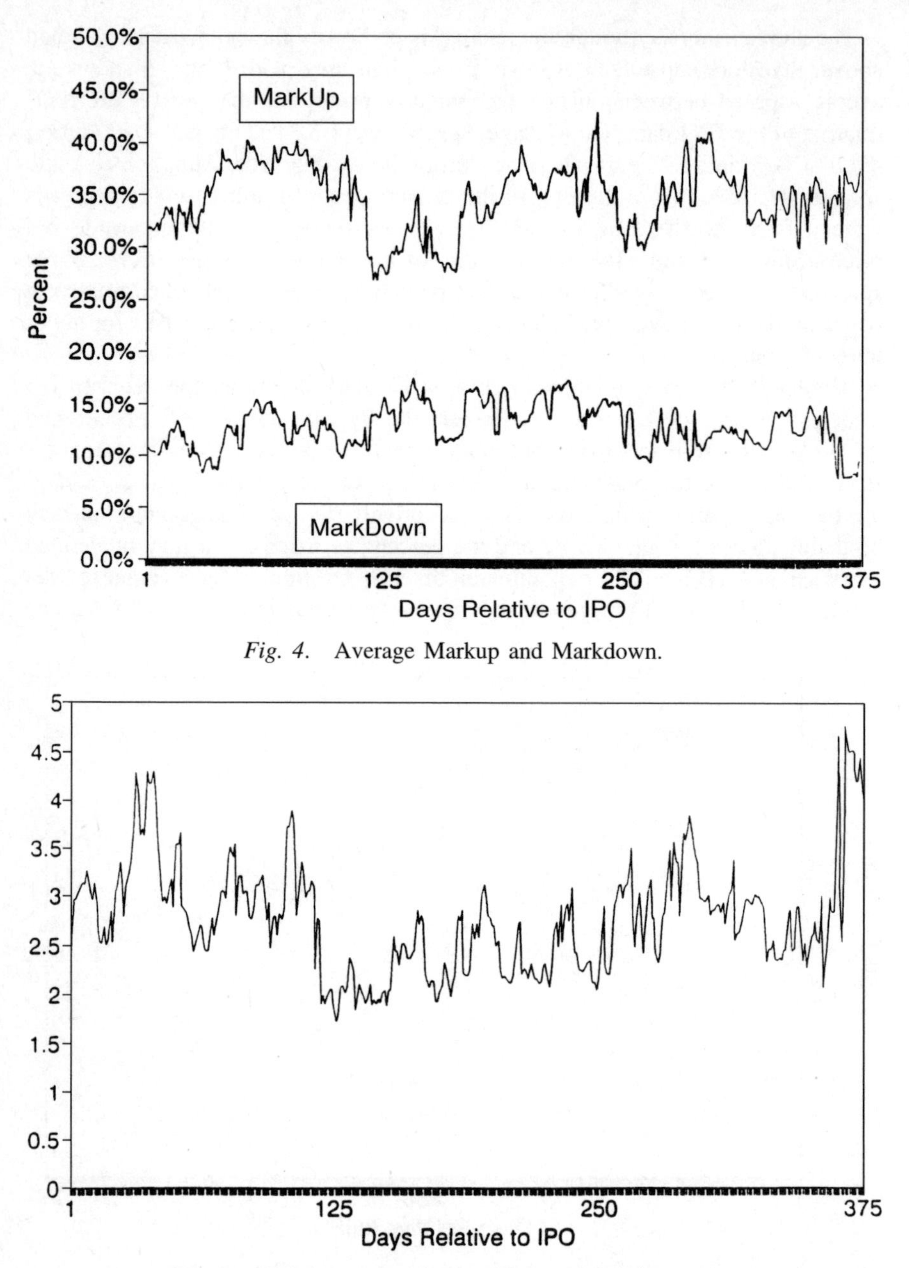

Fig. 4. Average Markup and Markdown.

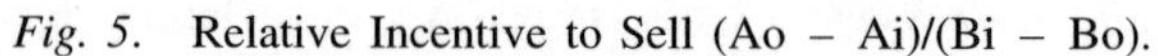

Fig. 5. Relative Incentive to Sell (Ao − Ai)/(Bi − Bo).

and the numerator is the difference between the wholesale (inside) and retail (outside) price quotes.

Fig. 4 plots the average mark-up and mark-down for the twelve BR IPOs during the first 375 days or year and a half of trading relative to the IPO date. Mark-ups averaged about 35% and mark-downs averaged about 13%. Thus, since broker compensation is related to mark-ups and mark-downs[13], there was clearly a greater incentive for broker to sell securities to customers rather than to purchase from customers. A more direct measure of the relative incentive for broker to sell rather than buy securities may be measure as the ratio of absolute mark-ups to absolute mark-downs $(A_o–A_i)/(B_i–B_o)$. This ratios is plotted in Fig. 5. On average over the first year and a half of trading broker had about 2.5 times more commission incentive to sell securities to customer than to purchase securities from customers.

The measure is useful to indicate the relative incentive given by BR under the normal circumstance where purchases and sales are independent and the broker's personal commission is essentially half of the difference between the inside price and the execution price. However, cross trades, due to even greater commissions were actively encouraged i.e. with a 50% share of commissions, a broker can potentially gain 0.5 (Outside Ask – Outside bid) in connection from cross trades.[14]

The practice of encouraging sales to customers and discouraging purchases from customers combined with a sales force of 1,800 brokers searching for new customers allowed BR to maximize its commission income and minimize the risk to market making in low priced securities. The combination of very risky securities and high trading costs meant that most investors in BR IPOs who purchased after the initial day of trading lose money.[15]

5. FURTHER DISCUSSIONS

Thus far, we have shown underwriters and corporate insiders benefited from the existence of the sub-standard security market. A critical question concerns the willingness of investors participate if they are invariably consistent losers. If simple irrationality is ruled out i.e. investors do not buy a stock simply because it is there, we have to assume that they are at least ex ante rational. Consequently, their participation in the market could be the result of committing one or more of the following mental and/or judgmental errors:

(a) They do not understand the adversary position of the broker/underwriter in this market. The broker/underwriter, as the dominant market maker, generates income from mark-ups and mark-downs on the securities traded and basically trades against outside investors; the greater the mark-up/down the higher the broker/underwriter's income and the lower the return to outside investors. Investors may not understand that advise given by brokers in such a situation should be taken cautiously if not skeptically. There is no free lunch.

(b) They lack the sophistication, training, time, etc., to incorporate known information and calculate value. Unfortunately, they are also willing to rely on biased sources.

(c) They succumb to the sales pitch (the three cold call approach)[16] of the brokers which are highly polished and much practiced.

(d) They could suffer from certain scale illusions or perceptions of securities and investments in general. These are:

(i) An illusion that a very small investment could bring a very large return. This illusion is similar to the illusions that cause large numbers of people to participate in lotteries even though it is known that lotteries are not "fair". The underwriter, by offering shares in the lowers dollar denominations (5 cents and under) and by massive splits of pre-offering shares (2,000 to 1 is typical), creates offerings of hundreds of millions of shares. Investors may be able to obtain several hundred thousand shares with a relatively small investment. This could create an illusion that "in case the company becomes the next Microsoft or Wal-Mart, one hundred thousand shares purchased year one for $1,000 could be worth over $10 million or even $100 million".

(ii) An illusion of scale on bid/ask differences. Investors might view a $20 spread an excessive ($30 bid/$50 ask) but view a $2 spread as less excessive ($3 bid/$5 ask) and be even less sensitive to a $0.20 spread ($0.30 bid/$0.50 ask). Finally, a $0.02 spread ($0.03 bid/$0.05 ask) could be viewed as "only two cents", thus grossly underestimating the magnitude of true transactions costs.

(e) Because the broker/underwriter is the sole source of price quotes and material information about the firm, they are in a position to manipulate expectations. Such manipulations might include spreading rumors of mergers, regulatory approval of products, large orders, joint ventures with reputable firms, etc. Other manipulations could involve the omission of unfavorable news/facts or even making untrue statements. Prior to the IPO, questionable practices could include steps taken to enable IPOs to be offered at inflated prices such as beefing up and window dressing past financial results, informally transmitting unsubstantiated favorable information and false over-reporting the extent of oversubscription of IPO shares. Subsequent to the IPO, questionable practices could include parking of shares, failure to reveal that the underwriter is the seller with unsubscribed shares and increasingthe mark-up to individual brokers to stimulate demand (rather than lowering prices). Deceptive trading practices could include cross-trades and by imposing a no net sales rule so that customers could only realize paper profits if realized funds are re-invested in other securities.

(f) Customers may fall prey to broker induced "beginner's luck" which may foster an unwarranted optimism concerning further investments. If a new customer has allocated some IPO shares at the offer price and is able to sell back on the first day, he/she can earn a high return. Consequently, the goodwill fostered by the profit in the first trade could induce a false optimism concerning further trades involving larger amounts of money, i.e. the Ponzi scheme.

Even if some investor behavior could be explained, it is still necessary to provide an explanation of why the supply of such naive investors is not soon exhausted, either through attrition (they lose all of their money) or from learning from experience. A NASDAQ report in September, 1989 by Scott Stapf estimated total losses to investors to be $2 billion a year out of $10 billion of activity. Obviously there has to be a continuously replenished supply of unsophisticated investors unearthed by heavy marketing efforts by brokers (who are known to place in excess of one hundred cold calls a day). Together, the ability of unsophisticated investors and the profits made by underwriters and insider in the sub-standard market is a social and economic concern i.e. mis-allocations of resources to welfare reducing investments with even more welfare reducing expenses to promote it.

The potential responses to the problems highlighted above are varied. Increased disclosure offers some promise but we show that currently available information, albeit prone to be biased, is quite adequate to assess the investments and to assign a much lower than the initial IPO prices. Basically, there is no amount of information that could help investors who lack the ability, training, time and willingness to use it. Other remedies include legal sanctions (driving the bad apples out of the business) and legislative means, e.g. the Penny Stock Reform Act of 1990 which imposed heavy disclosure requirements (the "Penny Stock Rule"), more detailed account statements, more information on compensation. The legislation has effectively choked off the market for stocks selling at $5 or less.

6. SUMMARY AND CONCLUSIONS

In this paper we examine in detail the characteristics of twelve IPOs brought to market by BR in the period from late 1985 through late 1987. In addition, we examine the return experience of investors subsequent to the IPOs and detail the microstructure of the penny stock market.

The topic is of theoretical interest because penny stock markets lack some of the conditions necessary for a standard efficient market such as the presence of large investors (institutions and individuals) and analysts following the securities. A study of such a market will provide a different set of laboratory parameters to study capital markets. Our study also provides a better understanding of the behavior and nature of small investors. It demonstrates that small investors are far more ill equipped to make intelligent investment decision, a sharp contrast to the more benign modeling of small investors in the literature, lacking clout or voice, e.g. Chemmanur and Fulghieri (1999), Mello and Parsons (1998), Stroughton and Zechner (1998).

The topic is of further interest because there has been an increase in the awareness of the role of marketing in finance. The introduction of new financially engineered products, offerings of new securities, closed-end funds, limited partnerships, etc. gives financial institutions greater role in marketing securities, in contrast to the familiar passive model of the financial intermediary. In the active model of security brokers, they are modeled as taking a mostly marketing role, i.e. identifying and contacting potential clients, explaining the nature of the securities, providing information, volunteering financial analysis, and estimating initial demand, etc. This study provides some insights as to how securities are "sold" (marketed) versus "bought" where the broker is merely an order taker and the investor conducts his/her own analysis. Our principal findings are:

1. No matter when outside investors invested in these firms, before, at, or after IPOs, they invariably lost most of their investment.
2. Contrary to expectations, the insiders/managers did pretty well in almost all cases despite of the fact that most of these firms ended up worthless, or near worthless.
3. BR did very well. It received incomes from several sources: underwriting fees, cheap stocks, consulting fees, as well as commissions/markup from making the market.
4. Items 2 and 3 could explain the supply of penny stock IPOs. The demand comes from the relatively larger sales force of Blinder that can reach tens of thousands of potential customers a day while only a few hundred or a couple of thousand of investors were needed for each IPO.
5. Evidence are also presented providing insights on how BR encouraged its broker to sell securities to clients rather than purchase form clients by using commission incentives to brokers.

In summary, this study provides an analysis of some practices that produce anomalous results in an important but neglected security market.

NOTES

1. BR & Co. was formed in New York State in 1977 and relocated to Colorado in 1977 and went on to become the largest penny stock brokerage firm in the United States. At its peak, BR employed 1,800 people in 82 branch offices in 35 states. It eventually collapsed

under the weight of litigation initiated by its allegedly defrauded clients and by federal and state securities authorities.

2. The SEC penny stock rules define a penny stock as one that has price of less than $5.00 per share and is not traded in a national securities exchange or the NASDAQ stock market. As of 1990 there were roughly 55,000 securities available for public trading in the United States, about 47,000 of which trade outside an exchange or NASDAQ.

3. One national publications (*Venture*) surveyed 78 IPOs of less than $1 in 1982 and found that 38 of them (45%) "had participants who were convicted felons, securities violators, targets of securities injunctions, reputed crime figures, or principals who face serious charges of insider financial misdealing". Barnes, 1983, p. 38.

4. Brokers had discretion to execute trades below the outside ask price (but above the inside ask) and above the outside bid price (but below the outside bid). Thus the mark-up and mark-down were somewhat negotiable. Broker compensation was based on mark-ups and mark-down so that large price concession would be unlikely since usually half of such concessions came out of the brokers pocket (see Petersen & Failkowski, 1994).

5. The Department of Banking and Finance of the State of Florida, (State – 1989) alleged that Blinder-Robinson manipulated the prices of certain of its IPOs, including World Wide Bingo. Bingo was first offered to the public on August 7, 1986. In Florida, 5 million shares were initially sold to 85 public customers at the offer price on $0.01 per share. The IPO was completely sold out on the first day. On the same day, BR began selling shares of Bingo in the secondary market to public customers at successively higher prices ranging from $0.025 to $0.0725. The Department alleged that purchasers in the secondary market were not informed that the shares they were purchasing had been sold by BR for $0.01 on the same day and had been purchased back from the initial customer for prices in excess of $0.01 and sold at a markup by BR. BR allegedly "repeated the process, commonly referred to as 'matched orders' or 'crossing' throughout August 7, and August 8, 1986, at successively higher prices. The Department finds that these higher prices resulted solely from manipulation of the market for these securities and were not the result of any news releases or announcements by World Wide Bingo, Inc. or any other event that would have prompted such an increase in the market value of the stock".

6. We use the term "commission" in a generic sense. Since these were principal trades (with BR as a principal) and not agency trades, the fees earned are better labelled mark-ups or mark-downs. BR and BR Brokers were compensated for executing two types of trades; normal mark-ups and mark-downs and cross trades. When a sale occurred from BR's own inventory the price would be marked-up above the BR inside ask price and the broker would be compensated as a percentage of the mark-up. Brokers could earn higher fees by executing matched or cross trades in which they would simultaneously find both a buyer and a seller for the same security. The State of Florida provides an example of the profitability of BR IPOs (State – page 57): "on August 7, 1986 , Broker X . . . sold 250,000 shares of Bingo to customer A at a price of $.01. For this transaction, respondent BR received gross commissions of 10% of the purchase price, or $250. That same day A sold the shares back to BR for $.025 per share, with respondent BR thereby receiving a gross commission of 50% of the price paid the customer A, $3,125 (of $6,250). This commission is derived from the fact that, while A was paid $0.025 . . . the shares were simultaneously sold to customers B,C, and D at a price of $0.05 per share ($12,500). On August 7, broker X began the day by selling 250,000 shares of Bingo at $.01. By the end of the same day, the shares were being sold for $0.05 . . . a price increase of 500%. As a result of the transactions by X that day, BR received gross commissions of $6,500".

The State went on to estimate that BR generated gross commissions from selling approximately 5 million shares of Bingo in Florida on August 7 and 8 alone of $135,840. This translates to roughly $4 million nationwide in two days for an original offering of 150 million shares with gross proceeds to the company of only $2.8 million.

7. It has been alleged that some BR brokers had spatulas on their desks which symbolized flipping or the practice of buying and selling IPOs on the first day at successively higher prices. Thus, investors who purchased at the offer would sell at a higher price and successive rounds (flips) would occur at successively higher prices. Thus, it seems likely that some investors in successive rounds in the first few days would also make money although it is difficult to quantify the amount.

8. These numbers are probably inflated somewhat from actual investor experience. Individual BR account executives had discretion to sell to customers essentially anywhere between the inside and outside ask prices. BR account executives were compensated, on average, at a rate equal to about one half of the difference between the inside and outside prices.

9. These figures show all prices available for BR until August 1988 when our data stopped. Subsequent to that time prices were obtained from various sources. When there are long discontinuities, linear price changes has been assumed.

10. In February 1987 Tele-Art Ltd founded Tel-Art Securities, a brokerage firm which was sold to BR four months later.

11. Note that prices on the right hand side of Figure 4 are less variable than those on the left side. This is due to, at least partially, a data problem where we had to obtain prices from various sources after August, 1988. IPOs occurred fairly uniformly throughout 1986 and 1987 so that relative to the IPO date there were fewer firms with good numbers (BR numbers) as the sample moved through time after the IPO date. Also, Pink Sheet prices are wholesale prices and thus are not strictly comparable with the BR retail prices used when available.

12. The Florida Department of Banking and Finance alleged that "for certain new issues in which respondent BR made a market, associated persons . . . were not allowed . . . to sell shares previously purchased by customers for at least ninety days unless such sales orders were crossed with purchase orders from other customers. When associated persons . . . tried . . . to sell the customers securities without a corresponding purchase order from another customer, the order was not accepted". (State – page 87).

13. Broker compensation is better viewed as related to absolute mark-up and mark-downs since the broker received about half of the difference between the execution price and the wholesale price as compensation. Over the period in question, the average absolute mark-up was about 1.9 cents and the average absolute mark-down was about 0.71 cents.

14. An example is in order. Suppose a penny stock has an inside (wholesale) bid-ask of four cents and six cents and an outside (retail) bid-ask of three and a half and eight cents. Suppose the broker buys and sells 10,000 shares to two different customers but does not cross the trades. His commission on the sale would be (10,000)(0.08–0.0.06)(0.5) = $150 and his commission on the purchase from the customer would be (10,000)(0.04–0.035)(0.5) = $25, for a total of $125. If the broker crossed the trades his commission would be (10,000)(0.08–0.0.035)(0.5) = $225.

15. It is not clear that investors who purchased at the offer price realized profits in aggregate. A reading of articles in the trade press on BR and various legal documents suggests that few investors ever got money out of BR once funds were committed. Paper

profits such as those which resulted from buying IPOs at the offer price were used as enticements to make further investments in other securities which were eventually consumed by large mark-ups and mark-downs.

16. The three call sales procedure works as follows: the first call from the broker to the investor is an introductory call to get acquainted; the second call is a courtesy call, and to add that if something interesting should come up, would he/she be interested; and the third call carry a sense of urgency to invest instantly due to unusual but fungible breaking news, etc.

REFERENCES

Barnes, E. (1983). Bad Pennies, *Venture*, November, 1983, pp. 38–46.

Chemmanur, T.J., & P. Fulghieri, (1999). A theory of going public decision, *Review of Financial Studies*, 249–279.

Fischel, D., (1982). Modern Finance Theory in Securities Fraud Cases Involving Actively Traded Securities, *The Business Lawyer*, *38*, 1–20.

Goldstein, J. I., & Cox, L. D. (1991). Penny Stock Markups and Markdowns, *Northwestern University Law Review*, 301–322.

Goldstein, J. I., Ramshaw, P. D., & Ackerson, S. B. (1992). An Investment Masquerade: A Descriptive Overview of Penny Stock Fraud and The Federal Securities Laws, *The Business Lawyer*, *47*, 773–835.

Guenther, D. A., & Willenborg, M. (1999). Capital gain tax and the cost of capital for small business: Evidence from the IPO market, *Journal of Financial Economics*, *53*(3), 385–408.

Hanley, K. W., Lee, C. M. C., & Seguin, P. J. (1993). The Marketing of Closed-End Fund IPOs: Evidence from Transaction Data, November, University of Michigan Working Paper.

Hansen, R., Fuller, B., & Jangigian, V. (1987). The Over-Allotment Option and Equity Floatation Costs: An Empirical Investigation, *Financial Management*, (Summer 1987) 24–32.

Lee, C., & Ready, M. (1992). Inferring Trade Direction from Intraday Data. *Journal of Finance*, *46*, 733–746.

Mello, A., & Parsons, J. F. (1998). Going public and the ownership structure of the firm. *Journal of Financial Economics*, 79–109.

Mitchell, M., & Netter, J. (1993). The Role of Financial Economics in Securities Fraud cases: Applications at the Securities and Exchange Commission. Mitsui Life Financial Research Center, Reprint Series No. 93–25, University of Michigan, 1993.

Monroe, M. A., & Seguin, J. (1991). An Evaluation of Penny Stocks, Mitsui Life Financial Research Center, Working Paper No. 91–16, University of Michigan, (March, 1991).

Petersen, M. A., & Failkowski, D. (1994). Posted Versus Effective Spreads: Good Prices or Bad Quotes? *Journal of Financial Economics*, *35*, June, 269–292.

Schultz, P., & Zaman, M. A. (1994). Aftermarket Support and Underpricing of Initial Public Offerings. *Journal of Financial Economics*, *35*, 199–220.

State of Florida, Department of Banking and Finance (1989). In the Matter of Blinder, Robinson & Co.) *Immediate Final Order to Cease and Desist*, Administrative Proceeding No. 1097-S-5/89, Docket May 17, 1989.2

Stroughton, N. M., & Zechner, J. (1998). IPO mechanisms, monitoring and ownership structure. *Journal of Financial Economics*, *49*(1), 45–77.

APPENDIX 1

A Brief Chronological History of the Blinder-Robinson Company.

March 1970 Incorporated in the State of New York. Meyer Blinder was the president, director and owner.

1971 Fine for inaccurate calculation of capital for broker. In subsequent years NASD (National Association of Securities Dealers) has filed the following complaints involving excessive markup, inaccurate or misleading information filed with NASD, improper handling of escrow funds from offerings, violation of NASD Rules of Fair Practice, etc. The complaints resulted in censures and fines.

June 1974 – NASD complaint No. NY-1842.
July 1977 – NASD complaint No. NY-SC-203.
July 1979 – NASD complaint No. D-436.
September 1986 – NASD complaint No. DEN-666.
February 1986 – NASD complaint No. MS-481-AWC.
September 1988 – NASD complaint No. MS-671
January 1989 – NASD complaint No. DEN-854

Also the company was the subject of the following civil and administrative actions by the SEC, relating to violation of federal securities laws:

March, 1978 – Civil complaint for injunction
August 1980 – Civil complaint for injunction
September 1982 – Affirmed NASD sanctions
April 1983 – Civil complaint for restraining order
December 1986 – Administrative complaint
1990 – Civil action No. 90-4523

By 1989 the company had 35 regulatory actions by 29 states which had resulted in fines, license denials, cease and desist orders, injunctions, suspension/revocation, and probationary and conditional registrations.

1984 The company hired John Cox as Vice-President. He was a former NASD director of antifraud section.

March 1985 Formed a holding company, Blinder International Enterprises, Inc., in Colorado. A stock offering was made in November 1986. The company later changed its name to Intercontinental Enterprises.

October 2, 1986	As of this date, Blinder-Robinson had 1,162 brokers in 45 offices over 18 states. At its peak, there were 1700–1800 brokers, 61 offices in 35 states. Each broker made 100 calls a day (Forbes, April 20, 1987).
June 1987	Blinder-Robinson bought the Tele-Art Securities (Hong Kong) from Tel-Art Inc.
March 23, 1987	Date Meyer Blinder was supposed to be banished from brokerage business for life. U.S. District Court granted him a stay.
1990	BR filed for Chapter 11 Bankruptcy.

APPENDIX 2

Brief Histories of Blinder-Robinson's IPOs in the Sample.

1. Allertech

The company was incorporated in Englewood, Colorado in November 1984. It was a specialty laboratory that marketed an easy to use combination of product and service in the allergy field to primary care physicians. The IPO was offered on April 4, 1986. From November 1984 to September 1985, it incurred losses every year with an accumulated total loss of $2,044,000 in comparison to total revenue of $704,000 in the same period. Later, it, as a shell corporation with listed stocks and was required by a private company, where its name was changed to Centerscope on May 18, 1990 and acquired by CW Electronics (CWE) a computer retailer in June 6, 1991. CWE made a $2,250,000 public offering of stock on July 15, 1993.

2. Amereco (AmerEco Environmental Services)

The company was incorporated in Missouri in June 1983 under its original name, "PCB Disposal Systems, Inc. Its business was in the management of toxic and non-hazardous waste, concentrating primarily in PCB contaminated fluids and materials. Amereco IPO was offered on April 3, 1987. The company was not profitable and filed for Chapter 11 on February 27, 1990.

3. HDL Communications

The company was founded in October 1985 as a vanity press which developed and published books in conjunction with the authors. It was incorporated in California originally as Merdyne Publishers, Inc., changed to Taliesin

Publishers, Inc., and finally to its present name in 1986. Its IPO was offered on March 3, 1987. In the next three years it suffered a cumulative loss of $1,830,000 and announced to discontinue business in June 1989. There were no revenue generating operations in 90-92, however, losses from expenses amounted to $300,000. The company is a listed shell company looking for possible (reverse) acquisition. There were two known unsuccessful merger attempts.

4. Pasta Via International, Inc.
The company's business was to develop a chain of specialty retail stores selling fresh pasta, sauces, and related product. It was established in November 1983 and offered its IPO on December 6, 1985. The new financing enabled the company to expand its operation, increasing sales from $120,000 in 1985 to over $1,400,000 in 1988 and 89 or a total revenue of $3,908,000 for the subsequent four years (1986–1989), versus total income for the same period of negative $4,200,000. In March 1989, the company had closed 10 of the 11 stores.

5. Paul's Place.
The company was incorporated in Colorado in January, 1985. It operated a chain of fast food restaurant. Its IPO was offered on November 12, 1986. The company reported total revenue of $3,665,000 and total loss of $4,915,000 for the post IPO years, 1987–1989. The original founder was bought out by a large shareholder in 1987. On October 1989 a private investor group headed by the same shareholder purchase the assets and assumed some liabilities. The remaining corporate shell still had negative equity.

6. Seafood, Inc.
The company, incorporated in Louisiana in 1978, operated a seafood distribution and processing business. The IPO was offered in July, 1987. The company suffered losses in subsequent years. The company closed all its plants and subsidiaries by mid 1989 and filed for bankruptcy (Chapter 11) on October 5, 1989.

7. Tekna-Tools.
The company, incorporated in Utah on October 25, 1983 manufactured and marketed small tools, such as multipurpose screw drivers. It changed its name from Pacific Western Tool, Inc., to Tekna-Tool (July 1984) to Nortek (August 1987) and back to Tekna Tool, Inc., (October 1987). The IPO was offered on December 7, 1987 for a net receipt of over $4 million. The tool business did not take off. Sales declined from $713,000 in 1988 to 292,000 in 1992, resulted in cumulated loss of $608,000. The company changed its name to Revotek in

April 1989. In 1991 it acquired and subsequently disposed of a bankrupt meat processing firm. Because the company did not use up all the cash received from IPOs, it had made several repurchases of its own shares and finally took the company private and liquidate with a 1 to 24,000 reverse split. The company had slightly over $2 million of cash on hand, roughly half of the IPO receipt at the time of liquidation in September 1992.

8. Tele-Art.

Tele-Art is a limited liability International Business Company registered in the British Virgin Islands. It was incorporated as a holding company for Tel-Art HK and its subsidiaries. The Hong Kong based company designed, developed, manufactured, and sold digital watches and other miniaturized electronic consumer products. It had production facilities in Hong Kong, China, and Europe (France and Ireland). Its IPO, offered in September 1986 for $5 million was the only unit offering in the sample. In February, 1987, Tele-Art, Ltd. found the Tel-Art Securities, a brokerage firm that was sold to BR four months later. Also, an officer of BR was listed on the management of Tele-Art. The company had made several acquisitions, (Carmen Fashion Furs) and invests in several companies and joint ventures. Overall, results had not been profitable in the two years after IPO. The company has not filed any SEC report since then, and there is virtually no market activity for the stock in the U.S. It is known to be still in operation in Hong Kong. The company, however, refused to provide information to inquiries from U.S. and over Hong Kong sources.

9. Thermacor.

The California company, incorporated in Nevada in September 1984 and offered its IPO on December 1985 was developing a line of portable thermal reduction devices designed to cool various portions of the body. From April 1987 to March 1993 the company reported a cumulated loss of $1,279,000. The company ceased operation in June, 1991 and moved to Utah in February 1992 by the new management team. In 1993, the company had practically liquidated, it had no asset. It attempted to reorganize earlier, with a known failed attempt to sell to a German firm. The promoters, in spite of the failure, was able to keep the majority of the shares as finder's fees and took control of the company. The company is a shell company looking for a (reverse) acquisition.

10. Trudy, Inc.

The company incorporated in Delaware in February 1987 was a successor to a Connecticut company, Norwest Manufacturing, which charged its name to Trudy Toys Co. and Trudy Corporation. It designed, manufactured and marketed plush stuff animals and other children items such as books and cassettes. The IPO was offered in July 14, 1987. Its cumulative loss from April 1987 to March 1992, with not a single profitable year, totaled $4,500,000 while sales declined from a peak of $7,347,000 in 1990 to $826,000 in 1992. It sold its main line of business. The company had been laying off employees and was in default of its bank loan.

11. Western Acceptance Corporation.

The company was incorporated in Nevada in May, 1985. Its business was to provide financing of non-standard insurance premiums for high risk and cash poor drivers. The company offered its IPO in September 17, 1987. From 1987 to 1992 it suffered losses in every year, totaling $5,308,000. However, due to its very high gross margin of 50% (interest rate charged versus cost of funds), it managed to obtain extra financing from various sources; convertible redeemable preferred stock, unit offerings, and Receivable Trust Certificate via securitization. It made unsuccessful forays into Missouri and Texas by acquiring several, mostly weak or failed companies via stock exchange. The company was deleted from NASDAQ in September 20, 1989 and was readmitted in July 30, 1990 after 1/100 reverse split. The original founder and president was replaced in 1989. The company's creditor, on July 1, 1993 declared it to be in default, and thus it was unable to conduct business due to lack of cash. The Ameritrust of Texas oversees processing and liquidation.

12. Worldwide Bingo

The company, incorporated in Colorado in May, 1986 marketed a broadcast bingo promotion package to radio and television stations. The IPO was offered on December 31, 1986. The company incurred loss of $529,000 in 1987. It announced an intention to restructure in October 11, 1988.

EXECUTIVE COMPENSATION AND EXECUTIVE CONTRIBUTIONS TO CORPORATE PACS

Kathleen A. Farrell, Philip L. Hersch and Jeffry M. Netter

ABSTRACT

This paper estimates the determinants of the contributions made by top executives to their firm's Political Action Committee (PAC). We find that executive's personal PAC contributions (proxy for the interest of the firm) are positively related to his shareholdings, income and option holdings (proxies for the interests of the executive). Contributions are also higher for CEOs and board members. This is direct evidence that the structure of the contracts between the firm and management, especially compensation, aligns manager's personal behavior with the interests of the firm.

INTRODUCTION

This paper analyzes the relationship between an executive's compensation, share ownership and positioning within the corporate hierarchy and his contributions to his firm's Political Action Committee (PAC).[1] The paper contributes to two very different literatures. First, the paper provides evidence on the effectiveness of executive compensation in tying the personal behavior of executives to the

Advances in Financial Economics, Volume 6, pages 39–56.
Copyright © 2001 by Elsevier Science B.V.
All rights of reproduction in any form reserved.
ISBN: 0-7623-0713-7

interests of their firms. Second, it provides direct evidence on the giving by important corporate officers to company PACs, which adds to our understanding of the role of firms and money in the political process.

We combine data on PAC contributions and executive compensation. We are thus able to analyze the relationship between direct personal actions of top executives (donating their own money to the firm's PAC) and an action of the firm. Thus, our study is unique in that we actually observe manager behavior and its relationship to the manager's stake in the firm – compensation and position. The only other literature that directly examines a manager's personal actions is the analyses of manager's stock trading. (see, for example, Seyhun, 1986, 2000). However, those studies tend to concentrate on whether the manager is more informed than the market, not the impact of the manager's actions on the firm. Other studies of managerial behavior usually examine a more indirect relationship, such as the effect of managerial compensation on firm performance.

The specific question we address is whether the variation in executives' personal PAC contributions is simply a function of the characteristics of the firm/industry in which the executive is employed, or is it systematically related to the executive's incentive compensation. If the latter is true, we believe the relationship between compensation and personal donations to a company's PAC provides interesting evidence on the agency relationship between top executives and their firm. Further, if the executive's donations to the PAC are correlated with other ways his personal behavior is tied to the interests of the firm, our results provide evidence on the effects of compensation in mitigating the agency problem.

Executive contributions have several impacts on the effectiveness of the firm's PAC. First, the executives' contributions add to the size of the PAC. Second, there is a corporate culture and signaling aspect to executive contributions. Because of their position in the corporate hierarchy, executive donations can set an example for lower level managers to contribute. Further, upper level executives may believe they have a responsibility to contribute to the company PAC. In part, this may be to set an example for lower level managers to contribute.[2] It is also likely that upper level executives identify their interests as closely related to the interests of the company.

We find that executives' PAC contributions are positively related to their, shareholdings, income and option holdings. In addition, directors and CEOs tend to donate more than other executives. Finally, there are impacts of firm and industry characteristics on executive donations. These results suggest that executive's personal behavior (in this case spending their own money) is related to his compensation and their position within the firm. We proceed by briefly reviewing the relevant literature on corporate donations to PACs and executive compensation.

BACKGROUND

Contributions to PACS

Access to government is an important consideration for many firms. For example, firms spend considerable sums on lobbying government officials and a recent study by Agrawal and Knoeber (2001) even suggests the importance of political factors in explaining the choice of corporate directors. The area of corporate political involvement that has received the most attention, both public and academic, is campaign finance.[3] To mount a serious campaign effort, most candidates for elective office require substantial campaign funds. Primary sources of this funding are interest groups, such as business and labor.

Under U.S. election law, however, a corporation can not directly contribute to the campaigns of individuals seeking federal office.[4] A firm is permitted, however, to use its general funds to establish and administer a corporate political action committee. All monies to be distributed by the PAC to political campaigns must come from *voluntary* contributions made to the PAC by the firm's professional and managerial employees and stockholders. [5]Individuals can contribute up to $5,000 per year to a PAC. PACs, in turn, are permitted to contribute $5,000 to a candidate per election. Thus if a candidate runs in both a primary and general election, he/she can receive $10,000 per election cycle. The $10,000 limit leads to the major weakness of the paper since it is a relatively small amount of money to these highly paid executives, and thus donations are a crude proxy for the relation between executive behaviour and the firm's interests. However, there is significant variation in executive donations and thus we believe our results are meaningful.

There are several reasons why firms may be motivated to support specific candidates for elective office. The firms can aid the election chances of candidates with whom they share similar views. In addition, contributions can lead to access to elected officials to influence the formulation of policies and legislation that affect their firm or industry. One unnamed executive said :

> The PAC gives you access. It makes you a player. These Congressmen, in particular are constantly fundraising. Their elections are very expensive . . . So they have an ongoing need for funds. It profits us in the sense to be able to provide some funds because in the provision of it you get to know people, you help them out. There's no real quid pro quo . . . But the PAC gives you access, puts you in the game (Clawson et al., 1992, p1).

There are several articles in the public choice and political science literature on the determinants of corporate PAC activity across different firms and industries (see for example Pittman, 1977; Zardkoohi, 1985; Master & Keim, 1985;

Grier, Munger & Roberts, 1994; Taylor, 1997). Neglected in this literature is an analysis of the determinants of individual executive donations, which is the focus of this paper. We concentrate on the relationship between executive compensation and executive donations.

Compensation and Managerial Incentives

One of the most important areas in corporate finance is the relationship between executive compensation, executive incentives, and firm performance. The basic agency problem, popularized by Berle and Means (1932) and modeled by Jensen and Meckling (1976) is that in a firm, as a result of the separation of ownership and control, the interests of the owners and the managers diverge. One way to try and reduce the agency problem of the separation of ownership and control is through the compensation plans of the managers (see Abowd & Kaplan, 1999; Murphy, 1999 for reviews of executive compensation literature). Our goal is to test whether the incentive effects of executive compensation are correlated with executives' donations to the company PAC (a proxy for the interests of the firm).

Most executive compensation packages contain four basic components: a base salary, an annual bonus, stock options and long-term incentive plans (including restricted stock plans) where incentive compensation is tied to measurable indicators of firm performance (Murphy, 1999). Technically speaking, any potential adjustments related to some measure of performance represent an incentive component. However, Murphy (1999) notes that the executive employment contract often includes a guaranteed minimum increase in base salary over the subsequent five-year contract period, which often mitigates incentive effects of the compensation scheme. Stock options provide a direct link between managerial rewards and stock-price appreciation. The payout from exercising options increases dollar for dollar with increases in stock price. However, they are not identical to stock ownership. For example, options sufficiently out-of-the-money and unlikely to be exercised, lose any incentive value (Murphy, 1999).

In addition to the explicit components of the executive compensation package, most executives own stock. Personal stock ownership, although typically insignificant as a percentage of total shares outstanding for the firm, often represents a significant component of an executive's wealth. Stock ownership provides the most direct link between shareholder wealth and CEO wealth. Stock ownership has both incentive and wealth effects for the executive, which complicates our analysis. We are interested in the relationship between the incentive effects represented by share and option ownership and an executive's willingness to directly help his firm with a personal donation to the PAC, however we note that wealth effects of stock ownership will also impact his PAC donations. In our empirical analysis we attempt to control for the wealth effects of stock ownership.

We also consider the executive's position within the firm as a determinant of contributions. Clawson et al. (1992) suggest there are two reasons top managers may participate at a high level in PAC contributions. First, they are integrated into the corporate culture and identify with the company. Second, upper level management is subject to coercion, non- contributions to the PAC may be held against them. We do not directly test these "corporate culture" arguments other than to suggest that firms attempt to set up the structure of management contracts to align manager's interests with those of the firm, and aligning interests is especially important for top management. Further, if a PAC is important there is a signaling component to the donations of top management. Therefore we attempt to control for managements' status within the firm in explaining their contributions.

DATA AND EMPIRICAL SPECIFICATION

We examine the relationship between executive compensation and other factors and executive PAC contributions using regression analysis. We discuss our data sources and empirical specification below.

Data

ExecuComp provides compensation data for each of the five highest paid executive officers of publicly held corporations in the S & P 500, the S & P Mid-Cap 400 and the S & P Small-Cap 600 and thus includes approximately 1500 firms. We begin with the firms followed by *ExecuComp* and limit our sample to those firms with a PAC contributing at least $10,000 to federal candidates during the 1995–96 election cycle. We choose 1995–1996 to include a presidential election year, a time of increased PAC contributions. We drop firms with less than $10,000 in PAC contributions, since they are relatively inactive in campaign finance.

Federal election law mandates that all contributions to a PAC of $200 or more be identified, by donor, in financial disclosure reports filed with the Federal Election Committee (FEC). The Center for Responsive Politics (CRP) uses the FEC data to maintain a web site that lists, for every corporate PAC, the amount of each separate donation to the PAC, the individual donor's name, the donor's employer, and the date of the donation. We cross-reference the CRP data with the *ExecuComp* sample to obtain PAC contributions by individual executive.[6]

If an executive was not with the same company in both 1995 and 1996 he was dropped from the sample.

In our regression analysis, the dependent variable is the combined amounts an executive contributed to his firm's PAC during 1995 and 1996. This variable is censored from both above and below. The maximum contribution permitted by law is $5,000 per year, thereby capping contributions at $10,000. The minimum observed value is zero. A value of zero, however, requires additional interpretation. Because the minimum recordable donation is $200, any donations smaller than this amount are reported as zero. Since $200 is relatively small this would appear to be a minor problem. However, many corporations allow executives to donate in either lump sums or through payroll deductions. In the latter case, an executive paid weekly could conceivably have $100 in donations deducted per week, for an annual total donation of $5,000. However, since each of these deductions would be below the $200 threshold, the total PAC donations would be recorded as zero. Unfortunately, we have no direct information on whether firms used payroll deduction, but we attempt to address the potential problem in two ways.

First, the FEC breaks out from total PAC contributions, the aggregate total from contributions of $200 or more. In the cases where the aggregate total is reported as zero (i.e. there are no instances of *any* individual contributing at least $200), we assume, that the firm has a weekly (or perhaps biweekly) payroll deduction plan in place. For example, Boeing pays its executives biweekly, but despite having a substantial PAC, had no reported contributions meeting the $200 threshold. These firms are dropped from the sample since essentially, there is an error of measurement in the executive donation variable. If the measurement error is random, coefficients for the explanatory variables in the regressions explaining executive contributions should be unbiased, but estimated with less precision. By eliminating firms with no reported donations of $200 or more, we hope to mitigate the measurement problem. Eliminating observations, however, possibly introduces sample selection bias. We assume the frequency of payroll payments adopted by a company is exogenous and eliminates the sample selection bias.

Second, for those firms with FEC records of individuals contributing the minimum $200, it is often possible to infer whether a payroll deduction plan was available. This can be accomplished by examining the pattern of donations reported by the CRP. For example, in a given year an executive could have twenty-four identical entries, occurring at biweekly intervals suggesting that payroll deduction was in effect. Therefore, we create a dummy variable, Payroll Deduction, that equals one, if *any* employee of the corporation (not necessarily one in the *ExecuComp* sample) was observed to have contributed to the PAC on an installment basis. If a payroll deduction plan is in place, the probability of observing PAC contributions from a given individual is diminished. We consider other possible effects of payroll deduction below.

Table 1 reports descriptive statistics on the executives in our sample. There are 1306 executives in the 302 firms in our sample. Five hundred and eighteen of these executives are observed to have contributed at least $200 to their corporate PAC. Thus, 39% of the executives contributed something, while most are observed to have contributed nothing.[7] The average recorded contribution by the 518 executives who contributed was $3690, with a minimum contribution of $200 and a maximum of $10,000 (49 contributed $10,000 – the legal limit). While these are not huge contributions (especially since the average income of the executives was $862,390) there is substantial variation in the contributions. The question we seek to address is whether this variation is simply a function of the characteristics of the firm/industry in which the executive is employed, or is it systematically related to the executive's incentive compensation.

Regressions Explaining Executive Contributions

We hypothesize executive PAC contributions are a function of an executive's financial incentives, financial capacity, status within the company, and the characteristics of his company and industry.

A significant component of an executive's relationship with his company is his shareholdings of the company's stock. An executive whose personal wealth is very sensitive to the performance of his company would be expected to contribute more heavily to his corporate PAC. We measure value of shares as the sum of both restricted and unrestricted shares held in 1995 multiplied by share price as reported by *ExecuComp*. Restricted shares are shares awarded to officers that are not registered under the 1933 Securities Act and thus cannot be transferred until certain requirements are met. A typical method by which restricted shares become registered is the executive holding the shares for a certain period of time, and thus could be forfeited on termination of employment. Therefore, restricted shares are not as valuable to the executive as comparable unrestricted shares, even though both are valued the same by *ExecuComp*.[8] Aside from any incentive effect, value of shares may also have a wealth effect on PAC contributions. The value of shares held is generally a major component of executive wealth. Consequently, if contributions were positively correlated with financial capacity, we would still expect a positive correlation between ownership and contributions, even if incentive effects of share ownership were absent. Ideally, to help isolate the two effects, a separate variable,

Table 1. Descriptive Statistics for Sample of Executives from Firms Followed by *ExeuComp* and where the Firms Contributed at Least $10,000 to a Federal PAC in 1995–1996.

	Mean	Standard Deviation	Minimum	Maximum
Percent of executives observed to have contributed to PAC	0.396	–	–	–
PAC Contributions	$1463.7	2626.5	$0	$10,000
PAC Contribution (Contributors Only)	$3690.2	3029.9	$200	$10,000
Income (in dollars)	862,390	766,635	$88,980	$8,774,700
Value of Shares	13,524,786	112,548,741	0	2,405,920,240
Value of Options (in dollars)	4,102,630	12,240,462	0	317,880,000
Value Exercisable Options (in dollars)	2,765,139	9,395,444	0	238,410,000
Value Unexercisable Options (in dollars)	1,337,491	3,704,593	0	79,470,000
Employees in firm	33,617	60,868	215	745,000
Percent of days the executive was a CEO in 1995–1996	0.223	0.403	0	1.0
Director, equals one if the executive was a director but not the CEO.	0.217	0.413	0	1.0

Note: The sample size is 1306 for all variables except PAC Contributions for contributors only where N=518.

measuring non firm related wealth would be included as a control variable. A larger coefficient on share-ownership wealth would then be a cleaner indicator of incentive effects. Unfortunately, such a measure of external wealth is unavailable, and we cannot directly separate the incentive from the pure wealth effects of owning shares. It seems reasonable, however, that given that all executives in the sample are presumably wealthy and the relatively small dollar amounts of contributions involved, we would not expect absolute wealth to be a major factor in PAC contributions. To allow for diminishing effects, value of shares held is entered in quadratic form.

Another component of financial capacity is, of course, income. We measure income by combining an executive's 1995 salary and annual bonus. Bonus is included as a measure of income because we are measuring PAC contributions over the 1995 and 1996 period and compensation during 1995. A bonus awarded in 1995 no longer provides incentives to the executive in 1996. Thus we argue that bonus is more closely related to income in this context.[9] As with value of shares, we enter income quadratically to allow for diminishing effects. Since income does not capture incentive effects, a positive effect of value of shares and an insignificant effect of income (at the observed wealth levels) would provide stronger support that value of shares is primarily an incentive motivator for an executive.

An additional measure of financial incentives (and wealth) is stock options. Executive options typically become vested (exercisable) over time. For example, many grants specify that 25% of the initial option grant becomes exercisable in each of four years following the grant. Executive options are non-tradable and are typically forfeited if the executive leaves the firm before vesting. Again, options held may capture a wealth effect in that currently exercisable options that are in-the-money embody wealth, in addition to an incentive effect. However, options that are out-of-the money or in-the-money but not exercisable provide primarily an incentive effect since they will only have value if stock price appreciates sufficiently or maintains the current level such that the options remain in-the-money. Options that are sufficiently out-of-the money may lose their incentive value if the executive perceives little chance of exercising.

To capture the effect of options on PAC contributions, we include the total value of outstanding options in two specifications of our model. In an attempt to isolate the pure incentive effect of options, we estimate an alternative specification of our model where the value of options is split into two components: the value of exercisable options which may include both a wealth and incentive effect and the value of unexercisable options which we would expect to capture purely an incentive effect. The value of options is calculated by *ExecuComp* as the immediate exercisable value of the options or in other words, the stock price less the exercise price. By definition, *ExecuComp* assigns a zero value to options that are currently out-of-the money. (*ExecuComp* does provide estimates of the Black-Scholes value of options granted in the current year but not for all options outstanding and thus we do not use this estimate.) Because of the incentive and wealth effects we expect the coefficients of all the options variables to be positive.

There is a signaling aspect to executive contributions. Although all of the executives in the sample are high ranking, the CEO and officers who also serve on the board of directors are the most visible and should have the greatest incentives to signal support for the political agenda of the firm. We define a variable, Percent CEO, as the percentage of the sample period in which an executive held the CEO position. We also create a dummy variable, Director, that equals one if an executive was not the CEO, but served on the board in either 1995 or 1996.[10] We expect the coefficients of both variables to be positive.

Executive rank and responsibility within the corporation are highly correlated with the financial compensation variables. Because we only control in a rough way for executive rank, through variables for CEO and board director, its possible that any positive effects attributed to the compensation variables could be due to a failure to completely control for executive stature. To remedy this possibility, we also run regressions restricting the sample to executives who held the CEO position throughout 1995 and 1996, the most homogenous component of the sample. Therefore, in the CEO only sample our findings for the impact of the compensation variables on contributions are due to the incentive (and financial capacity) effects of the compensation.

The contributions by an executive to his corporate PAC are expected to depend on the corporation's need for such funding. Firms actively engaged in the political arena, would be more apt to actively solicit funds as well. One indicator of active solicitation would be the adoption of payroll deduction for PAC contributions, which is captured by the payroll deduction variable discussed earlier. The overall effect of that variable is therefore ambiguous; the presence of a payroll plan increases the probability that an executive would feel compelled to make a contribution, but reduce the probability of the contribution being recorded in the FEC database.

There is empirical evidence that size of the firm, measured by number of employees, and the size of its PAC are positively related (e.g. Zardkhooi, 1985). In part, size of firm may be a prime determinant of a firm's involvement in politics.[11] If so, executives of larger firms could be urged to contribute more to their PACs. The positive correlation, however, may simply be due to larger firms having a larger pool of managers from which to solicit funds. In this latter case, perhaps because of free-ridership, the *average* contribution made by a firm's executives could be negatively correlated with the number of employees. To control for either possibility, we include number of employees as an explanatory variable.

Finally, firms in different industries have different incentives to interact with the political process. Grier, Munger and Roberts (1994), for example, found factors such as industry sales to government, import penetration, industry concentration, and regulation to positively contribute to PAC contributions by industry. To control for these factors we allow for industry fixed effects by including forty-one 2-digit SIC industry dummy variables in the analysis.[12]

After eliminating observations with missing data, the final sample contains 1306 executives from 302 companies in 42 industries. Because of the double censoring of the dependent variable, regression equations were estimated by Tobit with both an upper and lower limit.

RESULTS

We report the results of regressions explaining top executive contributions to their firm's PAC in Tables 2 and 3. Table 2 contains the results for the full sample of executives and Table 3 reports the results for CEOs only. In Table 2, columns 1 and 3 report equations excluding the industry fixed effects, while they are included in the equations in columns 2 and 4. We are not able to estimate the fixed effects for the CEO only sample due to the small number of observations. A likelihood test reveals the fixed effects to have explanatory power for the full sample, but their inclusion/exclusion has only minimal impact on the significance levels of the other explanatory variables. Consequently, we do not expect the omission of fixed effects to materially affect the results of the CEO only sample. We turn attention first to the results from the full sample basing our discussion on the fixed-effects models – equations 2 and 4 in Table 2.

The coefficient of value of shares held is positive and the squared value of shares coefficient is negative. The net effect, evaluated at all values in the sample, however is positive, indicating that share ownership increases PAC contributions, but at a diminishing rate. This is consistent with the incentive effects of share ownership more closely aligning the interests of the executive and the firm. We cannot at this point, however, rule out the possibility that we are observing simply the wealth effects of share ownership. The plausibility of wealth effects being present is strengthened by the net positive effect of income on PAC contributions, again at a diminishing rate. Income may, however, be serving as a proxy for executive responsibility; a point we return to when we discuss the results from the CEO sub-sample.

In equation 2 the value of all options held by the executive has a positive and significant coefficient. Since options can also have both an incentive and a wealth effect, this is the result we would expect. In equation 4, we attempt to separate wealth from incentives by decomposing value of options into its components, value of exercisable options and value of unexercisable options. The latter is presumed to capture only incentive effects. Neither coefficient, however, is significant. To check whether multicollinearity is the problem, we

Table 2. Tobit Equations where the Dependent Variable Equals PAC Contributions made by All Executives During the 1995–96 Election Cycle. The Dependent Variable is Censored from Below at 0 and from Above at $10,000. The Coefficients are Reported with the Standard Errors in Parentheses.

	(1)	(2)	(3)	(4)
Constant	−4786[a] (397.2)	−3610[a] (801.6)	−4786[a] (397.2)	−3583[a] (802.0)
Income = salary + bonus (in $1000s)	2.14[a] (0.519)	1.77[a] (0.519)	2.14[a] (0.520)	1.74[a] (0.521)
Income-Squared	−0.000246[b] (0.0000989)	−0.000245[b] (0.0000963)	−0.0000245[b] (0.0000992)	−0.000208[b] (0.0000964)
Value of Shares, includes restricted shares (in $1000s)	0.0238[a] (0.00491)	0.0230[a] (0.00480)	0.0238[a] (0.00491)	0.0230[a] (0.00480)
Value of Shares Squared	-1.08×10^{-8}[a] (2.47×10^{-9})	-1.08×10^{-8}[a] (2.47×10^{-9})	-1.20×10^{-8}[a] (2.91×10^{-9})	-1.01×10^{-8}[a] (2.40×10^{-9})
Value of Options (sum of value of exercisable and unexercisable options in $1000s)	0.0312[c] (0.0170)	0.0325[c] (0.0167)	– –	– –
Value Exercisable Options (in $1000s)	– –	– –	0.0288 (0.0260)	0.0209 (0.0248)
Value Unexercisable Options (In $1000s)	– –	– –	0.0381 (.0591)	0.0677 (.0583)
Employees (number of firm employees in 1000s)	7.33[a] (2.73)	12.6[a] (3.12)	7.31[a] (2.73)	12.6[a] (3.11)
Percent of days executive was a CEO in 1995–1996	1770[a] (509.0)	1999[a] (494.0)	1770[a] (509.0)	2000[a] (494.8)
Director equals one if the executive was director but not the CEO	1298[b] (437.8)	1227[a] (423.1)	1298[a] (437.8)	1228[a] (423.0)
Payroll Deduction, equals one if the firm had a payroll deduction for PACs	2622[a] (383.3)	2355[a] (385.1)	2625[a] (384.2)	2365[a] (385.3)
Fixed Effects, define dummy variables representing two digit SIC codes	No	Yes	No	Yes

Table 2. (continued)

	(1)	(2)	(3)	(4)
Log Likelihood	−5144.6	−5067.9	−5144.6	−5067.7
N	1306	1306	1306	1306
Degrees of freedom	1296	1255	1295	1254

Notes:
a denotes significant at the 1% level.
b denotes significant at the 5% level.
c denotes significant at the 10% level.

entered exercisable and unexercisable options into the equations separately. The coefficients of both variables (not reported) are positive and significant at the 10% level, indicating that multicolliearity is a problem and prevents us from isolating a pure incentive effect of options.

Other evidence provided in Table 2 suggests the importance of the executive's position in the firm as a determinant of the amount he donates to the company PAC. The coefficients on both the number of days the executive was a firm's CEO and the director dummy are positive and significant. Thus, executives who are at the top of the corporate hierarchy find it more important to personally donate to the PAC than do lower level executives. These are important results, because they indicate that those executives most responsible for setting corporate policy or culture are more willing to set an example through personal contributions to implement the corporate policy. This in turn is evidence that firms are able to devise mechanisms (compensation, and culture for example) that align the personal actions of higher level executives with what is best for the firm.

The coefficients on number of employees and the payroll deduction variables are also positive and significant. Number of employees (a measure of firm size) is a determinant of the importance of PAC politics to larger firms. The establishment of a payroll deduction plan for contributions is an indicator of the importance of maintaining a PAC to the firm (Clawson et al., (1992) using anecdotal evidence argue it is the most important indicator).

The positive coefficients on the number of employees and payroll deduction variables indicate that individual executives respond positively to their firms' needs for PAC contributions. Further, the positive coefficient on number of employees suggests that if free-ridership among executives is present, it does not fully negate the motivation to contribute.

Table 3. Tobit Equations where the Dependent Variable Equals PAC Contributions made by the CEO During the 1995-96 Election Cycle. The Dependent Variable is Censored from Below at 0 and from Above at $10,000. The Coefficients are Reported with the Standard Errors in Parentheses.

	(1)	(2)
Constant	−2260[b]	−2187[b]
	(1107)	(1105)
Income = salary + bonus (in $1000s)	1.04	0.934
	(0.960)	(0.965)
Income – Squared	-5.47×10^{-7}	-3.98×10^{-5}
	(1.52×10^{-4})	(1.53×10^{-4})
Value of Shares includes restricted shares (in $1000s)	0.0292[a]	0.0290[a]
	(.00764)	(.00759)
Value of Shares Squared	-1.29×10^{-8}[a]	-1.26×10^{-8}[a]
	(3.66×10^{-9})	(3.61×10^{-9})
Value of Options = sum Exercisible and Unexercisable options (in $1000s)	0.0138	–
	(0.0224)	–
Value Exercisable Options (in $1000s)	–	0.0125
		(.0368)
Value Unexercisable Options (in $1000s)	–	0.0915
		(0.0894)
Employees (number of firm employees in 1000s)	16.5[b]	16.1[b]
	(7.37)	(7.30)
Payroll Deduction, equals one if the firm had a payroll deduction plan for PACs	2307[a]	2332[a]
	(956.3)	(953.3)
Fixed Effects, define dummy variables representing two digit SIC codes	No	No
Log Likelihood	−1232.1	−1231.8
Number of observations	241	241

a denotes significant at the 1% level.
b denotes significant at the 5% level.
c denotes significant at the 10% level.

To quantify how the explanatory variables actually affect executive contributions we provide the following estimates based on the results from the second column of Table 2. Unlike ordinary least squares, Tobit coefficients do not measure the full marginal effects of the explanatory variables, but need to be adjusted by a scaling factor.[13] Based on the adjusted coefficients, the elasticity of PAC contributions with respect to value of shares held (evaluated at the means) is 0.079. Thus, as the value of executive share holdings increases by 1%, contributions increase by 0.79%. Income elasticity is 0.308, value-of-options elasticity is 0.034, and the elasticity for number of employees is 0.107. In addition, if the executive was the CEO for the full period add $741 to the donations, if he was a director (but not CEO) add $455, and if the firm had payroll deduction add $873.

In sum, all the variables that represent wealth and incentive effects are positively related to executive donations. The results are consistent with the hypothesis that compensation motivates executives to act in ways that are in the firm's best interest (especially if giving to the firm's PAC is correlated with other ways the executives act in the interest of the firm). Given the positive effect of income, however, we can not yet rule out that the relationship is solely a consequence of financial capacity.

The results from the CEO sample provide additional insights on the determinants of executive contributions to PACs. The purpose of examining a CEO only sample is that allows us to control for status within the corporate hierarchy. These results are reported in Table 3. The most interesting result is that the income coefficients are now insignificant, but the coefficients for value of shares held maintain their significance at the one-percent level.[14] Results therefore suggest that, at least for CEOs, the positive effect of shares held on PAC contributions is of an incentive nature and not of financial capacity. Further, the positive influence of income in the full sample may not reflect financial capacity. Instead income may be acting as a proxy for corporate status, and the responsibilities and expectations placed on higher level executives.

The coefficients of value of options held (and its components) are all insignificant in the CEO sample. Thus, for CEOs, we find neither an incentive nor wealth effect associated with options. Therefore, since we would expect the incentive effects of options to be the same or greater as shares held, we are cautious in attributing the effects of shares held solely to the incentives they confer.

CONCLUSIONS

This paper estimates the determinants of the contributions made by top executives to their firm's Political Action Committee (PAC). We suggest that examining the amount of money an executive donates to his/her firm's PAC and how it is related to variables that proxy for the executive's incentives and responsibility in the firm, provides evidence on the agency relationship inherent in the separation of ownership and control. Most importantly, we examine the relationship between executive compensation and executive position and executive donations.

The results suggest that compensation, position within the firm, size of the firm, industry, and the importance of PACs to the firm all have an impact on the executive's donations. We find that executive's personal PAC contributions are positively related to his shareholdings, income and option holdings. The contributions are also higher for higher-level executives and greater when the PAC is more important to the firm.

We suggest that these are important results in the study of how the structure of contracts between a firm and its top managers control the agency problem in the relationship. All these results are consistent with the hypothesis that executive's personal behavior (in this case spending their own money) is tied to what is in the best interest of the firm by compensation and his position within the firm. We find that executive compensation (a proxy for the interests of the executive) is related to the executive's contributions to the firm's PAC (a proxy for the interest of the firm). This is direct evidence that the structure of the contracts between the firm and management, especially compensation, aligns manager's personal behavior with the interests of the firm.

NOTES

1. For convenience, we use the masculine pronoun for executives.
2. Technically, contributions are supposed to be anonymous, but an executive can make a public display of his contribution.
3. Some good general references on the subject of money in politics include Clawson, Neustadtl and Scott (1992) and Sorauf (1988, 1992).
4. Corporations are permitted to give directly to political parties for "get out the vote campaigns" or general party building. These donations are generally referred to as "soft money" contributions. Since the parties can contribute to individual campaigns and because campaign funds are fungible, this is viewed by many as a loophole in campaign finance law. Corporations can also run "issue ads" that favor one candidate over another, but must act independently of the candidate's election committee.
5. Clawson et al. (1992) note that the employee contributions to a PAC are not as voluntary as say a contribution to a Sierra club PAC contribution. They use anecdotal evidence to illustrate the ways firms attempt to induce contributions from their employees.
6. We do not attempt to include contributions from spouses or other family members who may be shareholders of the firm. In part, this is due to the difficulty of accurately tracking spouse contributions (e.g. the possible use of their maiden names) and that

some executives may be unmarried. Additionally, since we employ a Tobit model with a $10,000 upper limit (the maximum allowed for an individual), spousal contributions are not a significant factor since it is likely that a spouse would only make an independent contribution if the executive wanted to contribute more than the $10,000 limit.

7. Clawson et al. (1992) using anecdotal evidence argue that the particiaption rate of top management in company PACs is close to 100%. Either their evidence is incorrect or we are missing a large number of very small contributions.

8. When we exclude restricted shares from the regressions the results do not change.

9. In some specifications, we used salary rather than salary plus bonus as our income measure. The overall results are unaffected. Entering salary and bonus separately introduces some multicollinearity problems, but does not change the basic results.

10. Director is defined to be mutually exclusive of Percent CEO, since virtually all CEOs also serve on the board. Separating the two avoids collinearity. Director is entered as a dummy variable, because *ExecuComp* does not provide the data on board tenure, which it does for CEO.

11. Watts and Zimmerman (1978) make this argument and present supporting empirical evidence. Agrawal and Knoeber (2000) find that larger firms are more likely to have directors with some background in politics.

12. Although it would be preferable to use firm-fixed effects instead of industry effects, the small number of observations per firm did not make this practical.

13. In Tobit, a change in an explanatory variable affects both the probability that the dependent variable will fall in the uncensored portion of the distribution and the conditional mean of the dependent variable in the uncensored portion of the distribution. Unless adjusted by a scalar transformation, the estimated Tobit coefficient vector reflects only the latter. See Greene (1997) for a discussion.

14. This result holds if income-squared is omitted as an explanatory variable.

REFERENCES

Abowd, J. M., & Kaplan, D. S. (1999). Executive Compensation: Six Questions That Need Answering. *Journal of Economic Perspectives*, *13*(4), 145–168.

Agrawal, A., & Knoeber, C. R. (2001). Do Some Outside Directors Play a Political Role? *Journal of Law and Economics* (forthcoming) *44*(1).

Berle, A. A. Jr., & Means, G. C. (1932). *The Modern Corporation and Private Property*. New York: Macmillan.

Clawson, D., Neusdtl, A., & Scott, D. (1992). *Money Talks: Corporate PACs and Political Influence*. New York: BasicBooks.

Greene, W. H. (1997). *Econometric Analysis 3rd ed.* Upper Saddle River, N.J.: Prentice Hall.

Grier, K. B., Munger, M. C., & Roberts, B. E. (1994). The Determinants of Industry Political Activity, 1978–1986. *American Political Science Review*, *88*(4), 911–926.

Herrnson, P. S. (1999). Financing the 1996 Congressional Elections. In: J. C. Green (Ed.), *Financing the 1996 Election*. Armonk, NY: M.E. Sharpe.

Jensen, M., & Meckling, W. H. (1976). Theory of the Firm: Managerial Behavior, Agency Costs and Ownership Structure. *Journal of Financial Economics*, *3*, 305–360.

Jensen, M., & Murphy, K. (1990). Performance Pay and Top-Management Incentives. *Journal of Political Economy*, *98*(2), 225–264.

Master, M. F., & Keim, G. D. (1985). Determinants of PAC Participation Among Large Corporations. *Journal of Politics*, *47*, 1159–1173.

Murphy, K. (1999). Executive Compensation. In: O. Ashenfelter & D. Card, *Handbook of Labor Economics*. New York: North Holland.

Pittman, R. (1977). Market Structure and Campaign Contributions. *Public Choice*, *31*, 71–80.

Seyhun, H. N. (1986). Insiders' Profits, Costs of Trading, and Market Efficiency. *Journal of Financial Economics*, *16*(2), 189–212.

Seyhun, H. N. (2000). *Investment Intelligence from Insider Trading*. Cambridge, MA.: MIT Press.

Sorauf, F. J. (1988). *Money in American Elections*. Glenview, IL: Scott, Foresman.

Sorauf, F. J. (1992). *Inside Campaign Finance*. New Haven: Yale University Press.

Taylor, D. (1997). The Relationship Between Firm Investment in Technological Innovation and Political Action. *Southern Economic Journal*, *63*(4), 888–903.

Watts, R., & Zimmerman, J. (1978). Toward a Positive Theory of the Determination of Accounting Standards. *Accounting Review*, *53*, 112–134.

Wayne, L. (1999). Following the Money, Through the Web. *New York Times*, August 26, 1999.

Zardkoohi, A. (1985). On the Political Participation of the Firm in the Electoral Process, *Southern Economic Journal*, *51*, 804–817.

IS MANAGERIAL EQUITY OWNERSHIP AN ALTERNATIVE GOVERNANCE MECHANISM FOR JAPANESE FIRMS?

Stephen P. Ferris, Kenneth A. Kim and Pattanaporn Kitsabunnarat

ABSTRACT

Due to the existence of keiretsu networks and influential bank shareholders, managerial-ownership is not viewed as important in Japan. With the recent decline in the power and influence of Japanese banks, this view might now be obsolete. We present evidence that managerial ownership has become an alternative mechanism for corporate governance in Japan. Using 1993 and 1996 data, we find that firms with significant managerial equity ownership are typically non-keiretsu firms and hold less bank debt. Further, these same manager-owned firms exhibit more control potential and make more discretionary expenditures than do other firms. Overall, our findings suggest that managerial equity ownership is a substitute governance mechanism for monitoring by banks and keiretsu.

Advances in Financial Economics, Volume 6, pages 57–81.
Copyright © 2001 by Elsevier Science B.V.
All rights of reproduction in any form reserved.
ISBN: 0-7623-0713-7

I. INTRODUCTION

Corporate business groups known as keiretsu and influential bank monitors are unique aspects of the Japanese corporate environment. Because of their existence, researchers argue that agency problems between Japanese managers and shareholders are minimal (Nakatani, 1984; Hoshi, Kashyap & Scharfstein, 1990, 1991; Prowse, 1990). Therefore, the need for an "owner-manager" to align interests between managers and shareholders has been viewed as an *unnecessary* governance mechanism for Japanese firms (Kang & Shivdasani, 1995; Kaplan, 1994; Prowse, 1992). Recent literature however contends that bank oversight might be ineffective in Japan (Kang & Stulz, 1998; Morck & Nakamura, 1999; Weinstein & Yafeh, 1998). These findings prompt Morck and Nakamura (1999) to contend that alternative forms of corporate governance must be operating, especially for non-keiretsu (independent) firms. This research is one of several recent studies that reconsiders the importance of owner-managers in Japan (Kang & Shivdasani, 1999; Morck, Nakurmura & Shivdasani, 1998). In contrast to prior research, we find that owner-managers serve as an alternative to the traditional devices for Japanese corporate governance.

Kang and Stulz (1998), Morck and Nakamura (1999) and Weinstein and Yafeh (1998) question the effectiveness of bank oversight in Japan. Morck and Nakamura (1999) argue that among independent firms, bank equity holders pursue their interests as creditors at the expense of their equity claims. These contentions are consistent with Weinstein and Yafeh (1998) and with the earlier suspicions of Aoki (1990) and Kester (1991). Further, since banks are primarily concerned with the firm's ability to meet its debt payments, banks are less involved with riskier firms (Nakatani, 1984; Weinstein & Yafeh, 1998). Gibson (1995) and Kang and Stulz (1998) also argue that poor bank health might adversely affect the investment prospects of their dependent firms. This particular contention is especially relevant to the late 1980s and early 1990s when Japanese banks experienced significant financial difficulties. In light of these findings, Morck and Nakamura (1999) contend that some independent firms might require corporate control mechanisms beyond that provided by bank oversight.

Given the recent research revealing the limitations of bank oversight in Japan, we re-evaluate the importance of Japanese owner-managers. Jensen and Meckling (1976) argue that managers holding large equity positions can align managerial interests with those of shareholders. The Jensen and Meckling agency view of the firm has received extensive empirical support with U.S. corporate data (Morck, Shleifer & Vishny, 1986; Cho, 1998; Himmelberg, Hubbard & Palia, 1998; Holderness, Krozner & Sheehan, 1999). International evidence however has only recently revealed the benefits of managerial-ownership (LaPorta, Lopez-De-Silanes & Shleifer, 1999). Previously, it was widely believed that other governance devices such as corporate or family conglomerates, bank control through equity, or various legal protections precluded the need to develop owner-managers as corporate monitors. These beliefs also explain why owner-managers were considered unimportant in Japan. But because the power of Japanese banks has declined, the increasing lack of a keiretsu affiliation for firms and the rarity of incentive-based managerial compensation, it is possible that managerial equity ownership now serves as an alternative governance mechanism.

Our initial analysis of Japanese equity ownership data reveals several interesting findings. From a sample of 1,053 non-financial Japanese firms in 1993, we identify 99 firms where a senior executive (chairman or president) is also one of the firm's five largest shareholders.[1] Of these 99 firms, 91 are non-keiretsu firms. Additionally, those firms with significant managerial equity ownership have lower levels of bank equity ownership. These observations tentatively suggest three non-mutually exclusive findings. First, the fact that senior executives are among the largest shareholders reveals that such an ownership structure might be useful for Japanese firms. Second, managerial equity ownership appears to be a possible alternative to keiretsu governance. Third, among independent firms, managerial equity ownership appears to be an alternative governance mechanism for bank control.

To further explore whether managerial equity ownership serves as an alternative corporate governance mechanism, we examine firm-specific characteristics such as firm risk (which proxies for a firm's control potential), firm age (to capture life-cycle effects), bank loans on the balance sheets (which reveals bank oversight and influence), and levels of discretionary expenditures (which also captures managerial control potential), to determine if they can predict the existence of a large-shareholder executive or explain variations in executive ownership levels.

Our findings provide consistent evidence that managerial equity ownership represents an alternative mechanism for corporate governance in Japan. Our results suggest that Japanese corporate governance is becoming similar to that of U.S. firms. This finding supports the general contention of Kaplan (1994) and Kang and Shivdasani (1995, 1996) that corporate governance in Japan is not so different from U.S. practices.

We organize the remainder of our study as follows. In Section II, we present various descriptive statistics for the variables used in our analysis. In Section III we discuss the various determinants of managerial equity ownership and provide an overview of our general research methodology. We offer our empirical findings in Section IV. In Section V we test for endogeneity in the relation between firm performance and the equity ownership structure of Japanese firms. We conclude in Section VI with a brief summary.

II. DESCRIPTIVE STATISTICS AND EQUITY OWNERSHIP IN JAPAN

We obtain the equity ownership data from issues of *The Japan Company Handbook* (1994, 1997) published by Toyo Kezai. We also obtain the age of the firm (Firm Age) from the *Handbook*. We collect financial statement and stock return data from the PACAP Databases-Japan.[2] We determine keiretsu membership from *Industrial Groupings in Japan* (1985, 1989) published by Dodwell Marketing Consultants. We cross check these keiretsu memberships with the more recent *Kigyo Keiretsu Souran* (1992) published by Toyo Keizai.[3] We limit this study to firms with complete data.

Our methodology involves the construction of two samples. The first sample is drawn from 1993 and allows a direct comparison with Kang and Stulz (1998). Our 1993 sample of 1,053 non-financial firms consists of 342 keiretsu-affiliated firms and 711 independent firms. A second sample is collected from 1996 and allows both an up-dated examination of the issue as well as providing an inter-temporal robustness test. Our 1996 sample contains 1,070 non-financial firms, with 342 keiretsu firms and 728 independent firms.

We use three different ownership measures in this study: (1) equity ownership concentration (Top5), (2) managerial equity ownership (Mgr-Own), and (3) bank equity ownership (Fin-Own). The ownership concentration measure, Top5, is calculated as the percentage held by the top five shareholders. Our use of the top *five* shareholders as a measure of ownership concentration results from three important considerations. First, as we include a greater number of shareholders, the ownership concentration measure becomes less informative as it eventually converges to 100%. Second, Demsetz and Lehn (1985) and Prowse (1992) establish the appropriateness of the top five shareholders' ownership as a measure of ownership concentration and consequent power. Finally, we use Top5 so that we can directly compare our findings to Prowse (1992) who uses a Top5 measure in his examination of the Japanese equity ownership structure.

For managerial ownership, Mgr-Own, we only consider the equity ownership held by the two highest-ranking executives. As Kang and Shivdasani (1995) and Kaplan (1994) document, the chairman and president are the Japanese managers who exert the most significant managerial control over the firm. We also include the firm's chairman as a senior executive since such individuals usually have powers equivalent to the president and because Japanese internal governance mechanisms are typically group rather than individual oriented (Kaplan, 1994). Similar to Prowse (1992), our measure for managerial equity ownership represents the percentage of outstanding shares held by the top executives when they rank among the top five shareholders.[4]

Finally, we examine the level of equity ownership held by financial institutions, Fin-Own, since financial institutions are viewed as influential shareholders in Japan. Therefore, we calculate the aggregate percent of outstanding shares held by commercial banks, credit banks, trust banks, securities companies, and insurance companies when they rank among the top five shareholders. Since 1987, commercial banks are prohibited from holding more than 5% of a firm's outstanding equity under the Anti-Monopoly Act of 1977. As a result, banks use affiliated financial institutions to assist in maintaining significant equity positions, and thus control over their dependent firms.[5] All variables, including ownership variables and other firm-specific variables that we examine in this study are described in Table 1.

Table 1. Description of Variables.

Top5	Percentage of shares held by the top five shareholders.
Mgr-Own	Percentage of shares held by the top executives (president and/or chairman) ranking among the top five shareholders.
Fin-Own	Percentage of shares held by financial institutions (commercial banks, credit banks, trust banks, securities companies, and insurance companies) ranking among the top five shareholders.
SE	Standard error of estimate from market model in which firm's monthly returns for a five-year period prior to, and including, the study-year are regressed on the monthly returns of market portfolio for the same period.
Bank Loan Ratio	Total bank loans divided by total assets at fiscal year-end.
Fixed Assets Ratio	Net fixed asset to total sales at fiscal year-end.
Capital Expenditure Ratio	Increase (decrease) in capital investment (net-fixed assets) plus (accounting) depreciation divided by net fixed assets at fiscal year-end.
Operating Income Ratio	Operating income to total sales at fiscal year-end.
Firm Age	Number of years since the firm's establishment year.
MVE	Market value of common equity (in millions of yen) at fiscal year-end.
Keiretsu Dummy	A dummy variable (1 = firm belongs to a keiretsu; 0 = otherwise).
Regulation Dummy	A dummy variable (1 = electric or transportation industry; 0 = otherwise).

Table 2 provides summary statistics for the variables described in Table 1. We divide our sample into two subsamples based upon the equity holdings of the senior executives. The first subsample consists of firms *with* a senior executive that is also one of the top five shareholders. The second subsample contains those firms whose senior executives are not one of the top five shareholders. Our

Table 2. Summary Statistics for Firms with Managers as Top 5 Shareholders versus Firms Without Managers as Top 5 Shareholders.

Panel A: Year 1993

	With Mgr-Own	Without Mgr-Own	t-statistic
Top5	28.631	32.703	−4.356***
	(8.207)	(13.584)	
Mgr-Own	7.138	0.000	
	(4.794)	(0.000)	
Fin-Own	10.628	16.200	−10.223***
	(4.941)	(6.937)	
SE	0.089	0.082	2.822***
	(0.023)	(0.022)	
Bank Loan Ratio	0.148	0.212	−3.380***
	(0.172)	(0.181)	
Fixed Assets Ratio	0.333	0.427	−2.672***
	(0.310)	(0.509)	
Capital Expenditure Ratio	0.120	0.085	0.951
	(0.104)	(1.096)	
Operating Income Ratio	0.045	0.032	1.748*
	(0.071)	(0.062)	
Firm Age	46.212	55.719	−5.532***
	(14.635)	(16.435)	
MVE	1451.524	2328.348	−3.099***
	(2052.001)	(5983.526)	
Number of Observations	99	954	

Table 2. Continued.

Panel B: Year 1996

	With Mgr-Own	Without Mgr-Own	t-statistic
Top5	31.081	32.234	−0.984
	(10.654)	(14.012)	
Mgr-Own	8.325	0.000	
	(6.378)	(0.000)	
Fin-Own	10.241	15.484	−8.662**
	(5.568)	(6.740)	
SE	0.078	0.070	4.081***
	(0.022)	(0.021)	
Bank Loan Ratio	0.166	0.212	−2.327**
	(0.177)	(0.183)	
Fixed Assets Ratio	0.330	0.394	−2.137**
	(0.241)	(0.514)	
Capital Expenditure Ratio	0.093	0.102	−0.423
	(0.134)	(0.546)	
Operating Income Ratio	0.059	0.043	2.168**
	(0.068)	(0.053)	
Firm Age	47.784	58.142	−5.361***
	(15.774)	(16.574)	
MVE	1307.806	2344.080	−3.068***
	(2421.228)	(7227.707)	
Number of Observations	97	973	

Notes: 1. This table shows the mean results for the variables used in the study. We differentiate between firms that have a senior executive as a top 5 shareholder (With Mgr-Own) and those that do not (Without Mgr-Own). Standard deviations are reported in parentheses. Panel A presents results for the year 1993 while Panel B shows results for the year 1996. Table 1 contains a description of the variables. t-statistics indicating the statistical significance of the sample differences are also reported.

2. ***,**, and * denote statistical significance at the 1, 5, and 10% levels, respectively.

criterion that a manager must be a top-five shareholder for a firm to be classified with significant managerial equity ownership is biased toward a finding of "no difference" between the two samples. Therefore, any differences between the two samples should be viewed as meaningful. We provide a t-statistic to measure the statistical significance of any difference between the two samples. Panel A presents our findings for 1993 while those for 1996 are contained in Panel B.

From Panel A in Table 2, we see that firms characterized by managerial ownership have lower equity ownership concentrations than firms without such managerial ownership (28.6% versus 32.7%, respectively). We also find that firms with managerial equity ownership have lower equityholdings by financial institution than firms without managerial ownership (10.6% versus 16.2%, respectively). These differences are statistically significant at the 1% level and suggest that some firms use managerial equity holdings as a component of their governance structure, while others rely on blockholders (such as financial institutions) or the keiretsu network to monitor their managers. For manager-owned firms, the manager owns an average of 7% of the firm. Considering that our sample firms are listed on the First Section of the Tokyo Stock Exchange, the 7% managerial equity-ownership represents a significant stake. These findings are consistent with the conjecture by Demsetz (1983, 1986) that the primary reason for a manager to concentrate holdings in one firm and thereby forego the advantages associated with diversification is to exert control over the firm. We obtain similar findings for 1996 in Panel B.

There are a number of other interesting differences between the subsamples reported in Table 2. The standard error of the estimate from the market model, which is a firm risk measure, is higher for firms with managerial equity ownership. Consistent with Demsetz and Lehn (1985), Prowse (1992), and more recently with Himmelberg, Hubbard and Palia (1998) and Holderness, Kroszner, and Sheehan (1999), we use the standard error as a proxy for control potential. Demsetz and Lehn (1985) argue that with increasing riskiness in the firm's environment, the gains to monitoring become greater. Therefore, a positive relationship between executive ownership and the standard error suggests a monitoring presence by manager-shareholders.

Table 2 also shows that firms with managerial equity ownership use less bank loans relative to firms lacking such ownership. This observation provides evidence that executive ownership substitutes for bank oversight. High bank loan levels suggests the existence of a bank monitor, consistent with Diamond (1984). Furthermore, Anderson and Makhija (1999), Kang and Shivdasani (1999), and Weinstein and Yafeh (1998) confirm that Japanese firms depending on bank monitoring maintain higher bank loan levels. Other noteworthy differences between our two subsamples are that manager-owned firms have less net fixed assets, more free cash flow (operating income ratio), and are younger (Age) and smaller (MVE).

III. DETERMINANTS OF MANAGERIAL EQUITY OWNERSHIP

If managerial equity ownership is a governance device, then its presence should not be random. Rather, we should expect to observe high levels of managerial ownership where the benefits from such oversight are likely to be the greatest. This implies that managerial equity ownership can be endogenously determined from firm-specific characteristics. Specifically, the firm characteristics that we examine are as follows:

Control Potential: According to Demsetz and Lehn (1985) and Grossman and Hart (1986), and Holderness, Kroszner, and Sheehan (1999), there are both costs and benefits to insider ownership. When the firm's operating environment is risky, the benefits of oversight via manager-shareholder alignment exceeds (1) the cost of external monitoring to the shareholder and, (2) the cost of holding a large personal stake in a risky security to the manager. When the riskiness of the firm's operating environment is high, the potential gains to active oversight (and the potential for control to be important) are greater. As a result, firm risk represents an important measure of a firm's control potential. Consistent with Demsetz and Lehn (1985), Prowse (1992), Himmelberg, Hubbard and Palia (1998) and Holderness, Kroszner and Sheehan (1999), we use the standard error from the market model in which the firm's monthly returns are regressed on the monthly returns of a market portfolio for the same period. For the 1993 (1996) data, this period includes 1989–1993 (1992–1996).

Bank Loans: High levels of bank loans reveal a monitoring presence by banks (Diamond, 1984). This is especially true in Japan where financial institutions "persuade" firms to hold more bank debt in exchange for their monitoring services (Weinstein & Yafeh, 1998). Anderson and Makhija (1999) and Kang and Shivdasani (1999) also use bank loans as a way to measure the magnitude of bank oversight provided to a Japanese firm. Additionally, Kaplan and Minton (1994) identify firms with large bank debt and find that bank officials are appointed as corporate directors if the firm performs poorly. Since bank loan levels indicate the intensity of bank monitoring and because managerial ownership and bank oversight might substitute for one another, this implies a negative relation between bank loans and managerial ownership. We use the ratio of total bank loans divided by total assets as our estimate for bank loans.

Capital Intensity: According to Himmelberg, Hubbard and Palia (1998), fixed capital is easily monitored by outsiders. Therefore, if fixed capital is important to the firm's inputs, then the optimal levels of managerial ownership is lower. We use the ratio of fixed assets to total sales to measure the extent that fixed assets are present in the firm's asset structure.

Scope for Discretionary Spending: According to Himmelberg, Hubbard and Palia (1998), firm expenditures that are discretionary are not easily monitored by outsiders. Therefore, if these expenditures are necessary to the firm's operations, then the optimal levels of managerial ownership is higher. Because high growth firms require more investment spending, we construct a proxy that measures the opportunities for discretionary spending: the ratio of capital expenditure to fixed assets (Capital Expenditure Ratio). Jensen (1986) notes that for firms with high levels of free cash flow, the optimal levels of managerial ownership is also high. We use the ratio of operating income to sales (Operating Income Ratio) as a proxy for free cash flow (Himmelberg, Hubbard & Palia, 1998).

Firm Age: Younger firms are more likely to be in the growth stage of their development. As a result, younger firms might require more active oversight due to the more frequent agency conflict that will likely exist between managers and shareholders. The existence of owner-managers represents one solution to this problem. Consequently, younger firms are more likely to have manager-shareholders who are also the firm's founders. Finally, firms founded during the post-war era are less likely to be bank or group oriented since they are less frequently affiliated with one of the pre-war Japanese zaibatsu (the pre-cursor of today's keiretsu). For these reasons, we expect that younger firms will have higher levels of managerial equity ownership.

Regulation: According to Demsetz and Lehn (1985), Holderness, Kroszner, and Sheehan (1999), and Kole and Lehn (1998), the optimal level of managerial equity ownership is lower for regulated firms because their activities are limited and already monitored by regulators. To incorporate the impact of regulation on managerial ownership levels, we use a dummy variable equal to one if the firm belongs to the transportation or electric industries. In addition, we employ dummy variables for all other industries.

Firm Size: The larger the firm, the greater is the cost of obtaining a given fraction of ownership. Therefore, a negative relation between firm size and ownership reveals a wealth constraint. We follow Demsetz and Lehn (1985) and Prowse (1992) and use the market value of equity (MVE) as our measure for firm size.

Keiretsu: Berglöf and Perotti (1994) and Prowse (1992) argue that the keiretsu form of organization already employs an inherent form of mutual monitoring that is based on a complex nexus of inter-corporate and main bank relationships. Thus, keiretsu firms are less likely to employ owner-managers as a governance device. We construct a dummy variable that is equal to one if the firm belongs to a keiretsu and zero otherwise.

Based on the preceding discussion, if managerial equity ownership is a governance device, then it should be a function of various firm-specific characteristics:

$$\text{Managerial ownership} = f(\underset{(+)}{\text{control potential}}, \underset{(-)}{\text{bank loans}}, \underset{(-)}{\text{capital intensity}},$$

$$\underset{(+)}{\text{discretionary spending}}, \underset{(-)}{\text{firm age}}, \underset{(-)}{\text{regulation}}, \underset{(-)}{\text{firm size}}, \underset{(-)}{\text{keiretsu affiliation}}),$$

where the signs in the parentheses indicate the hypothesized relation between the independent variables and managerial equity ownership. To test this model, we conduct both Logit and Tobit regressions. For the Logit regressions, the dependent variable is equal to one if a manager is one of the five largest shareholders of the firm and zero otherwise. For the Tobit regressions, the dependent variable is the percentage of the outstanding equity held by the owner-manager.

IV. EMPIRICAL RESULTS

A. The Demsetz and Lehn Two-Factor Model

Before presenting the results from an analysis with the full model, we conduct a preliminary investigation using the Demsetz and Lehn (1985) two-factor model of ownership structure. The Demsetz and Lehn (1985) model specifies the firm's control potential and size as its independent variables. Using this model for Japanese data, Prowse (1992) finds that: (1) large shareholders monitor *non*-keiretsu firms and, (2) that financial institutions are the important large shareholders. Prowse (1992) however uses 1984 data, which precedes the emergence of the problems that plagued the Japanese banking industry later in the decade. Because this analysis examines 1993 and 1996 data and hence represent periods surrounding the Japanese banking crisis, we anticipate different results from Prowse. To conduct our preliminary regression analysis, we examine only non-keiretsu firms where firm risk (SE) and firm size (MVE) are the independent variables. We use Top5, Mgr-Own, and Fin-Own as the dependent variables in three separate Tobit regressions.[6] We present our results in Table 3.

Panel A contains our results for 1993 results while those for 1996 are contained in Panel B. The key finding in this analysis is the positive relationship between Mgr-Own and control potential. In the absence of a keiretsu affiliation and in the presence of a high level of control potential, it appears beneficial for senior executives to hold significant equity positions in the firm. Such holdings allow executives to exercise effective control and to maintain

Table 3. Tobit Regression Results of Equity Ownership: Two-Factor Models.

Panel A: Year 1993

Explanatory variables	Dependent Variables		
	Top5	Mgr-Own	Fin-Own
Intercept	30.339	−22.704	17.748
	(202.346)***	(33.776)***	(273.923)***
SE	31.060	99.148	−38.082
	(1.558)	(7.570)***	(9.260)***
MVE	−3.790	−39.720	−5.000
	(0.237)	(1.821)	(1.640)
Number of Observations	711	711	711
Log Likelihood	−2870.099	−531.359	2365.475

Panel B: Year 1996

Explanatory variables	Dependent Variables		
	Top5	Mgr-Own	Fin-Own
Intercept	26.896	−27.973	17.365
	(261.779)***	(40.120)***	(373.312)***
SE	83.579	131.282	−46.604
	(10.788)***	(8.610)***	(14.888)***
MVE	−2.579	−43.930	−2.630
	(0.154)	(1.793)	(0.715)
Number of Observations	728	728	728
Log Likelihood	−2962.510	−535.303	2365.475

Notes: 1. This table shows Tobit regression estimates where the dependent variables are the percentage of outstanding shares held by the top five shareholders (Top5), by the senior managers (Mgr-Own) and by financial institutions (Fin-Own) that rank among the top five shareholders. The samples include all non-keiretsu and non-financial Tokyo Stock Exchange listed firms with complete data during 1993 and 1996. Panel A presents results using 1993 data while Panel B shows results for 1996. Coefficients on MVE should be multiplied by 10^{-5}. Chi-square statistics are reported in parenthesis. The models' number of observations and log-likelihood statistics are also reported.

2. *** denote statistical significance at the 1% level.

sufficient power over the firm. We also see that Fin-Own is not positively related to control potential for either year. In fact, Fin-Own and control potential are negatively related, which suggests that financial institutions are risk-averse shareholders. This might explain the need for an alternative governance mechanism.

The positive relation between Mgr-Own and control potential, and the negative relationship between Fin-Own and control potential, contradicts the findings of Prowse (1992). Prowse finds financial institutions to be the primary monitors for non-keiretsu firms.

B. Additional Evidence: Full Model Results

We now provide the regression results from our fully specified model. By considering these additional variables, we achieve two objectives. First, we are able to confirm the relation between managerial ownership and control potential in the presence of control variables. Second, by considering bank loan levels, and other firm specific characteristics that specifically require managerial monitoring, we are able to test the contention that: (1) executive ownership represents an alternative governance mechanism and, (2) executive ownership substitutes for bank oversight. Further, based on firm specific characteristics, we should be able to predict which firms will employ managerial equity ownership as a governance device. Thus, we conduct a Logit analysis with Table 4 containing our results.

From the findings in Table 4, we see that firms with higher idiosyncratic risk levels are more likely to have a manager serve as a large shareholder. This finding is also consistent with the 2-factor Tobit analysis. Further, we find that firms with higher bank loan levels are less likely to have managers as one of their top five shareholders.[7] For the discretionary spending variables (capital expenditure and operating income ratios), only the operating income ratio (which proxies free cash flow) is significant in predicting the presence of an owner-manager. Since discretionary expenditures are not easily monitored, there are advantages to managerial equity ownership in firms where such expenditures are high.

We find that younger firms are more likely to have owner-managers. Agency conflict between managers and shareholders might be more significant for younger firms since they are more likely to face crucial decisions in terms of project selection, dividend payout and financing alternatives. Thus, younger firms might decide to use owner-managers as a way to alleviate these conflicts. We also recognize that younger firms are more likely to have their founders serve as owner-managers, which suggests that managerial equity ownership

Table 4. Logit Estimates of the Likelihood of Having a Manager Among the Top 5 Shareholders.

Explanatory Variables	Year 1993	Year 1996
Intercept	0.401 (0.139)	0.437 (0.185)
SE	13.087 (7.220)***	10.972 (4.640)**
Bank Loan Ratio	−1.439 (3.601)*	−0.816 (1.207)
Fixed Assets Ratio	−0.351 (0.784)	−0.391 (0.784)
Capital Expenditure Ratio	0.019 (0.007)	−0.159 (0.694)
Operating Income Ratio	4.150 (4.175)**	6.487 (7.806)***
Firm Age	−0.032 (13.443)***	−0.029 (11.753)***
Regulation Dummy	−1.178 (1.217)	−1.252 (1.191)
LogMVE	−0.245 (4.163)**	−0.291 (5.991)**
Keiretsu Dummy	−1.571 (16.716)***	−1.249 (11.466)***
Industry Dummies	Yes	Yes
Number of Observations	1053	1070
Pseudo-R^2	0.143	0.144

Notes: 1. This table shows Logit regression estimates of the probability that a firm has a manager among the top 5 shareholders. The sample includes all non-financial Tokyo Stock Exchange listed firms with complete data during 1993 (first column of results) and 1996 (second column of results). See Table 1 for a description of the explanatory variables. Industry dummy variables are based on Tokyo Stock Exchange classifications. Chi-square statistics are reported in parentheses. The models' number of observations and pseudo-R^2 are also reported.
2. ***, **, and * denote statistical significance at the 1, 5, and 10% levels, respectively.

might not be the result of an optimization process. Finally, regulated firms, larger firms, and keiretsu firms are all less likely to have an owner-manager.

In Table 5 we report our Tobit regression results. The explanatory variables remain identical to those used in the Logit analyses, but the dependent variable is the ownership percent held by managers. In summary, the Tobit regression results are consistent with the Logit results. Again, as predicted, firm control potential and the level of managerial equity ownership are positively related. As the control potential for a firm increases, so too does the extent of managerial equity ownership. The level of bank loans indicates the degree of the bank's involvement with the firm. This variable is negatively related to managerial equity ownership levels, which suggests a substitution between oversight by bank-shareholders and that by manager-shareholders. Further, we find that firms experience greater levels of managerial equity ownership as the magnitude of free cashflow increases. Finally, we conclude as before that younger and smaller firms are associated with higher levels of managerial equity ownership.

We also consider the possibility of a non-linear relation between our independent variables and managerial equity ownership. To test for this possibility, we include both a linear and a quadratic form of the independent variables in our models. None of the quadratic forms of the variables were statistically significant. We conclude that there is no evidence to suggest non-linearities in our hypothesized relations.

C. Evidence on Younger Firms

In the preceding section, we find that younger firms are more likely to have an owner-manager and that firm age and managerial equity ownership levels are negatively related. Such a result might be due to the characteristically high level of agency conflict present in younger firms. This finding however might be due to an increased tendency for younger firms to have founders that are large equity holders. This suggests that a firm's level of managerial equity ownership is not the result of a optimization process.

To test for such a possibility, we conduct an additional set of empirical investigations. First, we identify the median year of founding for all of our manager-owned firms. This year is determined to be 1948. 363 firms in our study sample were founded after 1948, but fewer than 50 of these firms have an owner-manager. Thus, most young firms do not have owner-managers. Further, we estimate various regression analyses on a sub-sample of firms that includes only young firms (i.e. firms founded after 1948). These regressions

Table 5. Tobit Estimates of Managerial Equity Ownership.

Explanatory Variables	Year 1993	Year 1996
Intercept	3.169	2.795
	(0.211)	(0.139)
SE	64.831	80.040
	(3.809)*	(3.878)**
Bank Loan Ratio	−10.070	−8.131
	(4.232)**	(2.040)
Fixed Assets Ratio	−2.377	−2.951
	(0.934)	(0.824)
Capital Expenditure Ratio	0.049	−1.241
	(0.001)	(0.682)
Operating Income Ratio	32.947	59.538
	(5.973)**	(10.902)***
Firm Age	−0.203	−0.223
	(13.194)***	(12.240)***
Regulation Dummy	−6.544	−7.145
	(1.359)	(1.106)
LogMVE	−1.754	−2.468
	(5.174)**	(7.851)***
Keiretsu Dummy	−9.349	−8.854
	(17.107)***	(11.958)***
Industry Dummies	Yes	Yes
Number of Observations	1053	1070
Log Likelihood	−570.805	−575.296

Notes: 1. This table shows Tobit regression estimates where the dependent variable is the percentage of outstanding shares held by senior managers (Mgr-Own) that rank among the top five shareholders. The sample includes all non-financial Tokyo Stock Exchange listed firms with complete data during 1993 (first column of results) and 1996 (second column of results). See Table 1 for a description of the explanatory variables. Industry dummy variables are based on Tokyo Stock Exchange classifications. Chi-square statistics are reported in parentheses. The models' number of observations and log-likelihood statistics are also reported.

2. ***, **, and * denote statistical significance at the 1, 5, and 10% levels, respectively.

attempt to explain the level of managerial equity ownership across these sub-samples. Tables 6 and 7 contain our findings.

For firms founded before 1948 (older firms) and for firms founded after 1948 (younger firms), the regression results are reported in the first and second columns, respectively. We find that for younger firms, greater managerial equity ownership is associated with more control potential and less bank debt. These results further verify that the degree of managerial equity ownership is the result of an optimization process. We conclude that managerial equity ownership can serve as a monitoring device for firms with control potential and as a substitution for bank-shareholder oversight. Finally, we again see that managerial equity ownership is primarily adopted by non-keiretsu firms, regardless of firm age. Overall, our examination of the firm's control potential, bank loans, discretionary spending and age yields further evidence that managerial ownership serves as an alternative mechanism for corporate governance in Japan.

V. ENDOGENEITY BETWEEN FIRM PERFORMANCE AND THE EQUITY OWNERSHIP STRUCTURE

Recent literature has tried to determine the direction of causality between the equity ownership structure and firm performance. Morck, Shleifer, and Vishny (1988) and McConnell and Servaes (1990) argue that managerial equity ownership can produce superior firm performance, but that firm value decreases at high levels of equity ownership because disciplinary takeovers become more expensive. In contrast, Demsetz and Lehn (1985) and more recently Cho (1998) and Himmelberg, Hubbard and Palia (1998) contend that the equity ownership structure is endogenous to firm specific characteristics.

To examine the issue of endogeneity between the firm's performance and its equity ownership structure, we use Cho's (1998) simultaneous systems model specification. Specifically, we model three regression equations in a simultaneous equations model.[8] In this framework, managerial-ownership is modeled as a function of firm performance, capital investment (capital expenditure ratio), the market value of common equity, firm risk, liquidity, and industry dummy variables. In another regression, firm performance is modeled as a function of managerial ownership, capital investment, bank loans, asset size and industry dummy variables. We use returns to the firm's common equity as our measure of firm performance.[9] Specifically, firm performance is measured as the compounded monthly stock returns cum-dividends during 1993. In the third regression specification, capital investment is modeled as a function of managerial equity ownership, firm performance, firm risk, liquidity, and a set of industry dummy variables. Table 8 presents the simultaneous-equations results.

Table 6. Logit Estimates of Managerial Equity Ownership: Older Firms versus Younger Firms.

Explanatory Variables	Old Firms	Young Firms
Intercept	0.114	0.231
	(0.005)	(0.018)
SE	7.386	16.959
	(1.113)	(5.298)**
Bank Loan Ratio	0.099	−2.615
	(0.008)	(4.986)**
Fixed Assets Ratio	−1.185	−0.010
	(2.417)	(0.000)
Capital Expenditure Ratio	0.029	−0.566
	(0.009)	(0.221)
Operating Income Ratio	9.536	2.253
	(7.851)***	(0.737)
Firm Age	−0.015	−0.035
	(1.248)	(2.495)
Regulation Dummy	−0.488	−12.329
	(0.199)	(0.001)
LogMVE	−0.351	−0.198
	(3.811)*	(1.339)
Keiretsu Dummy	−1.493	−1.727
	(9.146)***	(7.785)***
Industry Dummies	Yes	Yes
Number of Observations	690	363
Pseudo-R^2	0.141	0.140

Notes: 1. This table shows the Logit regression estimates of the probability that a firm has a manager among the top 5 shareholders during 1993. The regression is executed separately on sub-samples, which are formed by firm age. Firms founded before and after the year 1948 are considered "Old Firms" and "Young Firms," respectively. See Table 1 for a description of the explanatory variables. Industry dummy variables are based on Tokyo Stock Exchange classifications. Chi-square statistics are reported in parentheses. The models' number of observations and pseudo-R^2 are also reported.
2. ***, **, and * denote statistical significance at the 1, 5, and 10% levels, respectively.

Table 7. Tobit Estimates of Managerial Equity Ownership: Older Firms versus Younger Firms.

Explanatory Variables	Old Firms	Young Firms
Intercept	0.047 (0.000)	5.918 (0.348)
SE	29.103 (0.310)	83.379 (3.633)*
Bank Loan Ratio	−1.174 (0.021)	−14.163 (5.113)**
Fixed Assets Ratio	−9.242 (2.556)	0.800 (0.084)
Capital Expenditure Ratio	0.075 (0.002)	−2.783 (0.148)
Operating Income Ratio	72.788 (8.328)***	13.203 (0.744)
Firm Age	−0.124 (1.659)	−0.246 (3.481)*
Regulation Dummy	−3.082 (0.191)	−60.297 (0.000)
LogMVE	−2.270 (3.230)*	−1.573 (2.520)
Keiretsu Dummy	−10.045 (8.911)***	−9.065 (8.503)***
Industry Dummies	Yes	Yes
Number of Observations	690	363
Log Likelihood	−270.920	−290.999

Notes: 1. This table shows Tobit regression estimates where the dependent variable is the percentage of outstanding shares held by senior managers (Mgr-Own) that rank among the top five shareholders during 1993. The regression is executed separately on sub-samples formed on the basis of firm age. Firms founded before and after the year 1948 are considered "Old Firms" and "Young Firms," respectively. See Table 1 for a description of the explanatory variables. Industry dummy variables are based on Tokyo Stock Exchange classifications. Chi-square statistics are reported in parentheses. The models' number of observations and log-likelihood statistics are also reported.
2. ***, **, and * denote statistical significance at the 1, 5, and 10% levels, respectively.

The key findings from Table 8 are that managerial ownership is *endogenous* to control potential and that managerial ownership does not enhance firm performance. We find evidence that managerial ownership is endogenous of firm performance which is consistent with Cho (1998). Further, we observe that the firm's capital expenditures are driven more by firm liquidity than the level of equity ownership held by the firm's managers.

To test the robustness of our results, we conduct several additional examinations. First, we test our simultaneous systems model where accounting profits, a Q-proxy, and an industry-adjusted Q replaces stock returns as the measure for firm performance. The findings from this analysis are qualitatively similar to our earlier findings and hence are not separately reported. Next, we estimate the simultaneous system using different ownership variables including Top5 and Fin-Own. Again, we find no statistically significant relation between ownership and firm performance.

Finally, we check for a non-monotonic relationship between firm performance and ownership structure. Morck, Shleifer and Vishny (1988) and McConnell and Servaes (1990) conduct piece-wise regressions and find that at certain levels of insider equity ownership there is a positive relationship between insider equity stakes and corporate value. At different levels of insider ownership, Morck, Shleifer, and Vishny report a negative relationship between insider equity holdings and corporate value. Stulz (1988) notes that at high levels of insider ownership, the cost of a disciplinary takeover is higher, thus leading to lower firm values. Because disciplinary takeovers in Japan are virtually non-existent (Kester, 1991) however, we do not anticipate a significant non-monotonic relationship between performance and the equity ownership structure. Our empirical tests confirm our expectations and hence we do not separately tabulate our findings. Morck, Nakamura, and Shivdasani (1998) report similar results.

VI. CONCLUSION

The earlier literature on Japanese corporate finance contends that the keiretsu organizational form and large bank shareholders provide the primary corporate oversight for the Japanese firm. Consequently, owner-managers in Japan were viewed as unimportant. The more recent literature however establishes that when banks experience financial difficulties, their ability to effectively monitor diminishes. Given the crisis in Japanese banking that extended from the late 1980s to the early 1990s, Morck and Nakamura (1999) argue that alternative forms of governance have developed in Japan. In this paper, we suggest

Table 8. Simultaneous Equations Model of Managerial Equity Ownership, Firm Performance and Investment.

	Dependent Variables		
Explanatory Variables	Mgr-Own	Stock Mkt. Returns	Capital Expenditure Ratio
Intercept	−162.158 (53.961)***	0.813 (12.134)***	−0.109 (−0.520)
Mgr-Own		0.004 (1.344)	0.003 (0.237)
Stock Mkt. Returns	134.288 (46.941)***		−0.074 (−0.497)
Capital Expenditure Ratio	0.276 (0.047)	0.043 (1.131)	
MVE	−18.351 (17.449)***		
SE	106.257 (11.317)***		0.441 (0.300)
Cash Flow Ratio	−47.979 (6.142)**		5.039 (5.612)***
Bank Loan Ratio		−0.247 (−6.095)***	
Log (Total Assets)		0.024 (4.268)***	
Keiretsu Dummy	−4.490 (4.502)**	−0.039 (−2.651)***	0.053 (0.762)
Industry Dummies	Yes	Yes	Yes
Number of Observations	1048	1048	1048
Adjusted R2		0.081	0.020
F-statistic		8.113***	2.738***
Log Likelihood	−556.587		

Notes: 1. This table shows estimates of a simultaneous equation analysis of managerial ownership (Mgr-Own), firm performance (Stock Market Returns), and investment (Capital Expenditure Ratio). For the Mgr-Own equation, tobit estimates are presented, for the Stock Market Returns and Capital Expenditure models, least-squares estimates are presented. See Table 1 for a description of the explanatory variables. Industry dummy variables are based on Tokyo Stock Exchange classifications. The coefficient on MVE should be multiplied by 10^{-4}. t-statistics are reported in parentheses (chi-square statistics for the Mgr-Own equation).
2. ***, **, and * denote statistical significance at the 1, 5, and 10% levels, respectively.

that managerial equity ownership serves as one of these alternative governance mechanisms. Using ownership data from 1993 and 1996, we provide empirical support for this contention by reporting the following observations:

1. firms with executives as top shareholders are typically non-keiretsu firms, suggesting that managerial equity ownership might substitute for keiretsu governance,
2. firms with significant managerial equity ownership have lower financial institution ownership, suggesting that managerial-ownership is a substitute for bank oversight,
3. managerial equity ownership and bank loan levels are negatively related, confirming the previously mentioned substitution of managerial equity ownership for bank monitoring,
4. managerial equity ownership and firm age is negatively correlated, suggesting that younger firms use managerial equity ownership as a way to mitigate agency conflict,
5. managerial equity ownership is significant when firm expenditures are more discretionary,
6. managerial ownership and a firm's control potential are positively related, which also suggests managerial monitoring,
7. and managerial ownership appears to be endogenously determined.

NOTES

1. The chairman and president are the managers that exert the most control over a Japanese firm and possess approximately equivalent powers (Kang & Shivdasani, 1995; Kaplan, 1994).

2. PACAP Databases are created in cooperation with Daiwa Institute of Research and Toyo Kezai, Inc. The Pacific-Basin Capital Markets Research Center (PACAP) at the University of Rhode Island represents the vendor for the database.

3. The keiretsu classification scheme that we adopt is that of Dodwell (1985, 1989). Their definition considers the group's influential power, which is measured by the ratio of the group members' shareholding to the total shares held by the top ten shareholders. In addition, the keiretsu classification is also based on the following factors: (1) the characteristics and historical background of the group and the company, (2) the different sources and amounts of bank loans, (3) whether or not board members come from other group companies, (4) the company's attitude toward the group, and (5) the company's connections to non-group companies and to other groups. See Weinstein and Yafeh (1995) for further discussions of this classification scheme.

4. It is possible that a senior executive could be a significant equity holder, but still not among the top five shareholders. Further, Morck, Shleifer and Vishny (1988) find that the alignment of interest between managers and shareholders can occur at low levels of managerial ownership. By neglecting low levels of managerial ownership however,

it becomes more difficult for us to demonstrate that such ownership is an effective governance device.

5. See Kim and Limpaphayom (1998) for more discussion in this regard.

6. Even when we use a sample that includes keiretsu firms and a keiretsu dummy variable, the results remain qualitatively the same. We restrict our sample to non-keiretsu firms so we can directly compare our results to Prowse (1992), who focuses on a sample of non-keiretsu firms.

7. Holderness, Krozner, and Sheehan (1999) use a leverage ratio rather than a bank loan ratio. Consistent with Jensen (1986), they argue that leverage is a disciplinary mechanism and should be negatively related to managerial ownership. Using total liabilities divided by total assets as a leverage measure, we also find a negative relation with managerial equity ownership. Because we are particularly interested in the substitution between managerial equity ownership and bank involvement however, we only report results using bank loans.

8. Our simultaneous system follows the procedures of Blundell and Smith (1986) which allows for an embedded Tobit model in a system with more than two equations.

9. Cho (1998) uses Q as the firm performance variable. We use stock returns however for two reasons. First, due to the significant negative relation between managerial ownership and debt, a Q-proxy (calculated as the market value of equity plus the book value of debt divided the book value of total assets) suffers from significant collinearity. Second, Morck, Nakamura, and Shivdasani (1998) indicate that accurate Japanese Q measures are difficult to obtain.

ACKNOWLEDGMENTS

The authors thank Brent Ambrose, Chris Anderson, Dan Deli, Jon Garfinkel, Stu Gillan, Tim Haas, Jun-Koo Kang, Dick Pettway, Anil Shivdasani, Jonathan Sokobin, Ehsan Soofi, Masahiro Yoshikawa, John Zhou, seminar participants at George Washington University, the University of Wisconsin-Milwaukee, the 1999 FMA meeting in Orlando, and the 2000 FMA European meeting in Edinburgh for useful discussions and/or comments on earlier versions of this paper. A part of this work was conducted while Kim was a Visiting Scholar with the U.S. Securities and Exchange Commission. The Securities and Exchange Commission, as a matter of policy, disclaims responsibility for any private publication or statement by any of its employees. The views expressed herein are those of the authors and do not necessarily reflect the views of the Commission or of the authors' colleagues upon the staff of the Commission.

REFERENCES

Anderson, C., & Makhija, A. (1999). Deregulation, disintermediation, and agency costs of debt: evidence from Japan. *Journal of Financial Economics*, *51*, 309–339.

Aoki, M. (1990). Toward an economic model of the Japanese firm. *Journal of Economic Literature*, *28*, 1–27.

Berglöf, E., & Perotti, E. (1994). The governance structure of the Japanese financial keiretsu. *Journal of Financial Economics*, *36*, 259–284.

Blundell, R., & Smith, R. (1986). An exogeneity test for a simultaneous equation tobit model with an application to labor supply. *Econometrica*, *54*, 679–685.

Cho, M., (1998). Ownership structure, investment, and the corporate value: an empirical analysis. *Journal of Financial Economics*, *47*, 103–121.

Demsetz, H. (1983). The structure of ownership and the theory of the firm. *Journal of Law and Economics*, *26*, 375–390.

Demsetz, H. (1986). Corporate control, insider trading, and rates of return. *American Economic Review*, *76*, 313–316.

Demsetz, H., & Lehn, K. (1985). The structure of corporate ownership: Causes and consequences. *Journal of Political Economy*, *93*, 115–117.

Diamond, D. (1984). Financial intermediation and delegated monitoring. *Review of Economic Studies*, *51*, 393–414.

Dodwell Marketing Consultant (1985, 1989). *Industrial Groupings in Japan*. Dodwell Marketing Consultants, Tokyo, Japan.

Gibson, M. S. (1995). Can bank health affect investment? Evidence from Japan. *Journal of Business*, *68*, 281–308.

Grossman, S., & Hart, O. (1986). The costs and benefits of ownership: A theory of vertical and lateral integration. *Journal of Political Economy*, *94*, 691–719.

Himmelberg, C. P., Hubbard, R. G., & Palia, D. (1998). Understanding the determinants of managerial ownership and the link between ownership and performance, *Journal of Financial Economics*, *53*, 353–384.

Holderness, C. G., Kroszner, R. S., & Sheehan, D. (1999). Were the good old days that good? Changes in managerial stock ownership since the Great Depression. *Journal of Finance*, *54*, 435–469.

Hoshi, T., Kashyap, A., & Scharfstein, D. (1990). The role of banks in reducing the costs of financial distress in Japan. *Journal of Financial Economics*, *27*, 67–88.

Hoshi, T., Kashyap, A., & Scharfstein, D. (1991). Corporate structure, liquidity, and investment: Evidence from Japanese industrial groups. *Quarterly Journal of Economics*, *106*, 33–60.

Jensen, M. C. (1986). Agency costs of free cash flow, corporate finance, and takeovers. *American Economic Review*, *76*, 323–329.

Jensen, M. C., & Meckling, W. (1976). The theory of the firm: Managerial behavior, agency costs, and ownership structure. *Journal of Financial Economics*, *3*, 305–360.

Kang, J. K., & Shivdasani, A. (1995). Firm performance, corporate governance, and top executive turnover in Japan. *Journal of Financial Economics*, *38*, 29–58.

Kang, J. K., & Shivdasani, A. (1996). Does the Japanese governance system enhance shareholder wealth? Evidence from the stock-price effects of top management turnover. *Review of Financial Studies*, *9*, 1061–1095.

Kang, J. K., & Shivdasani, A. (1999). Alternative mechanisms for corporate governance in Japan: an analysis of independent and bank-affiliated firms. *Pacific-Basin Finance Journal*, *7*, 1–22.

Kang, J. K., & Stulz, R. (1998). Do banking shocks affect borrowing firm performance? An analysis of the Japanese experience. *Journal of Business*, forthcoming.

Kaplan, S. N. (1994). Top executive rewards and firm performance: A comparison of Japan and the United States. *Journal of Political Economy*, *102*, 510–546.

Kaplan, S. N., & Minton, B.A. (1994). Appointments of outsiders to Japanese boards: Determinants and implications for managers. *Journal of Financial Economics*, *36*, 225–258.

Kester, C. W. (1991). *Japanese Takeovers: The Global Contest for Corporate Control*. Cambridge, MA: Harvard Business School Press.

Kim, K. A., & Limpaphayom, P. (1998). A test of the two-tier corporate governance structure: The case of Japanese keiretsu. *Journal of Financial Research, 21*, 37–51.

Kole, S. R., & Lehn, K. (1998). Deregulation and the adoption of governance structure: The case of the U.S. airline industry. *Journal of Financial Economics, 52*, 79–117.

LaPorta, R., Lopez-De-Silanes, F., & Shleifer, A. (1999). Corporate ownership around the world. *Journal of Finance, 54*, 471–517.

McConnell J. J., & Servaes, H., (1990). Additional evidence on equity ownership and corporate value. *Journal of Financial Economics, 27*, 595–612.

Morck, R. & Nakamura, M. (1999). Banks and corporate control in Japan. *Journal of Finance, 54*, 319–339.

Morck, R., Nakamura, M., & Shivdasani, A. (1998). Banks, ownership structure, and firm value in Japan. *Journal of Business*, forthcoming.

Morck, R., Shleifer, A., & Vishny, R. (1988). Management ownership and market valuation: An empirical analysis. *Journal of Financial Economics, 20*, 293–315.

Nakatani, I. (1984). The economic role of corporate financial grouping. In: M. Aoki (Ed.), *Economic Analysis of the Japanese Firm*. New York: Elsevier.

Prowse, S. D. (1990). Institutional investment patterns and corporate financial behaviour in the U.S. and Japan. *Journal of Financial Economics, 27*, 43–66.

Prowse, S. D. (1992). The structure of corporate ownership in Japan. *Journal of Finance, 47*, 1121–1140.

Stulz, R. (1988). Managerial control of voting rights: Financing policies and the market for corporate control. *Journal of Financial Economics, 20*, 25–54.

Toyo Keizai (1992). *kigyo keiretsu souran*. Toyo Keizai, Tokyo, Japan.

Toyo Keizai (1994). *Japan Company Handbook*. Toyo Keizai, Tokyo, Japan.

Toyo Keizai (1997). *Japan Company Handbook*. Toyo Keizai, Tokyo, Japan.

Weinstein, D. E., & Yafeh, Y. (1995). Japan's corporate groups: Collusive or competitive? An empirical investigation of keiretsu behavior. *Journal of Industrial Economics, 43*, 359–376.

Weinstein, D. E., & Yafeh, Y. (1998). On the cost of a bank centered financial system: Evidence from the changing main bank relations in Japan. *Journal of Finance, 53*, 635–672.

THE INFLUENCE OF MANAGERIAL REPUTATION ON DIVIDEND SMOOTHING

Kathleen P. Fuller

ABSTRACT

This paper develops and empirically tests a dynamic model of dividend smoothing and signaling. Intertemporal dividend smoothing results from a "signal jamming" problem whereby the manager uses dividends not only to convey information about future cash flows but also to influence perceptions of his abilities. Any excess cash left after the dividend is paid must be carried into the future at a dissipative cost, and any cash shortfall arising from promising too high a dividend must be borrowed at a cost. Since these costs are higher for low-ability managers, high-ability managers promise a lower dividend when future cash flows are high and a higher dividend when future cash flows are low. Thus, high-ability managers not only credibly signal their type and the firm's future cash flows, but also smooth dividends more than low-ability managers. Consistent with the model, I find that firms that smooth more have higher future values, larger price appreciations to unexpected dividend increases, and smaller price declines to unexpected dividend decreases. Further, I find that high-ability managers smooth dividends more than low-ability managers.

Advances in Financial Economics, Volume 6, pages 83–115.
Copyright © 2001 by Elsevier Science B.V.
All rights of reproduction in any form reserved.
ISBN: 0-7623-0713-7

I. INTRODUCTION

Lintner (1956) first proposed that firms smooth changes in their dividends relative to changes in their earnings. He found that firms changed their dividends only if there were substantial changes in managers' expectations of firms' future earnings. Fama and Babiak (1968) and Laub (1976) empirically document Lintner's smoothing phenomenon for a large cross-section of firms. However, even though researchers have long struggled to theoretically explain this phenomenon, dividend smoothing remains a puzzle. The unexplained stylized facts regarding intertemporal dividend policy are as follows:

- The "desired" dividend payout is a constant fraction of earnings and these payouts partially adjust toward the desired payout from the level of the previous payout.[1]
- Dividend payouts are less volatile than earnings.[2]
- Dividend increases occur more often than dividend decreases.
- Firms often maintain a specific dividend payment for numerous quarters between payout policy changes.

This paper's primary contribution is to explain dividend smoothing by deriving a model in which the equilibrium solution to a game with asymmetric information has managers smoothing dividends relative to earnings. Additionally, the equilibrium is consistent with the previously mentioned stylized facts and generates testable predictions beyond Lintner's dividend smoothing model. Finally, this paper provides a link between static dividend signaling and intertemporal dividend smoothing models.

I begin by analyzing a two-period model in which there are high- and low-ability managers of firms with varying future cash flows. The manager's ability is not known by the market but is known by the manager. All managers have an investment that will produce cash flows for the next two periods, and each manager announces a dividend to be paid at the end of each period. Further, it is assumed that the market can observe realized cash flows that are either paid out in the form of a dividend or kept in the firm as excess cash. As indicated by Jensen (1986), excess cash introduces a free-cash-flow moral hazard problem as managers may frivolously spend this money on negative net present value projects. I capture this by assuming firms must pay a carrying cost for any unpaid cash flow and that this cost is higher for the low-ability manager.

The model predicts that when the manager expects higher future cash flows, he would like to signal this by increasing the dividend level. However, if both low- and high-ability managers increase dividends, then the firm's market value will be the expected value of all firms in the market. Therefore, high-ability managers will actually pay lower dividends that signals both their ability and the firms' future cash flows. When the expected future cash flows are low, managers would like to keep dividends high to avoid signaling low earning prospects to the market. However, high dividends with low cash flows necessitate distress financing which is costly for both managers, though more so for the low-ability managers. Therefore, high-ability managers forego some value by promising a higher dividend and borrowing money. By accessing the financial markets, the manager signals his type to the market. Smoothing of dividends emerges as high-ability managers pay lower dividends when cash flows are high and higher dividends when cash flows are low than do low-ability managers. Further, dividends are also smoothed relative to earnings.

The model generates the following predictions:

- High-ability managers will smooth dividends more than low-ability managers.
- Firms with smaller unexpected dividend increases have larger price increases.
- Firms that smooth dividends more have larger price appreciations to unexpected dividend increases and smaller price declines to unexpected dividend decreases.
- Firms that smooth more have higher market values.

The theory is then confronted with data on a sample of firms with unexpected dividend changes between 1989 and 1993. The extent of dividend smoothing is estimated as the ratio of the standard deviation of quarterly dividends to the standard deviation of quarterly earnings. First, I test whether firms with more dividend smoothing have larger (smaller) price appreciations (depreciations) to unexpected dividend increases (decreases). Second, I verify that firms that have smoothed more in the past will have higher future values. Using both the percentage change in the market-to-book ratio and the percentage change in the market value of equity plus debt as the change in firm value, I find that firms that smooth dividends more do have greater positive changes in value than those that smooth less. Third, I find that high-ability managers smooth dividends more than low-ability managers, as predicted by the model.

Many theoretical papers researching dividend policy have concentrated on the information content of dividends.[3] These papers build on the assumption that dividends signal news regarding the firm's future earning potential. Similar to these papers the model assumes that managers have superior information than the market regarding the future earnings potential of the firm. However, the manager's reputational concerns also influences the dividend decision.

Compared to the literature on the informational content of dividends, existing work on dividend smoothing is sparse. Kumar (1988) develops a two-period signaling model where managers with private information about their productivity choose dividends to signal this information to the market. Since the model has only one dividend payment, cross-sectional dividend smoothing arises, i.e. differences in dividend levels are smoothed across unobservable types rather than over time. Fudenberg and Tirole (1995) show dividend smoothing by managers is an outcome of the optimal contract between the manager and the firm. In the model managers report their unverifiable earnings each period. For job preservation, managers have an incentive to report higher earnings in bad times and lower earnings in good times. Thus, dividends can convey information not present in the earnings report, so managers smooth dividends so no information contradictory to their reported earnings exists.

Chowdhry and Nanda (1994) model the decision between stock repurchases and cash dividends. Though repurchases are cheaper due to the tax advantage, if shares are not significantly undervalued, the premium necessary to repurchase shares may be greater than the tax costs of dividends. Their optimal disbursement policy is to payout some cash as dividends and carry the remainder until the stock price is significantly undervalued. Then, the firm will use the excess cash to repurchase shares and cut the dividend payment. Warther (1996) rationalizes dividend smoothing as an equilibrium in an asymmetric information game between shareholders and management. Dividends discipline management since the failure to pay dividends may cause shareholders to realize the manager is "bad" and fire him. Only the worst manager will pay a low dividend and all other managers will pool at a higher dividend payout level. Thus, a smoothed dividend policy exists. Juster (1996) reconciles dividend smoothing and signaling in a dynamic model. There exists short-lived asymmetric information between the market and the manager. Due to higher taxes on dividends and the cost of raising outside equity, firms will pool and pay the same dividend level. Even if the costs of dividends are not high enough to ensure pooling, they will be enough that the dividend boost will be relatively small. Thus, the variance of dividends will be relatively small.

This paper takes a different approach. The focus is on the role of the managerial reputation in determining the future dividend levels. In this model, the firm's dividend payment is driven by the expected future earnings and the desire for the managers to signal their ability. Due to the asymmetric information present, a signal jamming problem arises as the dividend change signals not only the future earnings of the firm but also managerial type. Therefore, high ability managers smooth dividends to signal their type. The resulting dividend smoothing model is novel in that it allows both intertemporal smoothing and signaling to coexist.

The paper proceeds as follows. The model development and equilibrium solution is presented Section II. Section III discusses the model's empirical predictions that are then tested in Section IV. Finally, Section V concludes. All proofs are in the Appendix.

II. MODEL

The model has three dates, dates $t = 0$, 1, and 2. I assume everyone is risk neutral and the riskless rate is zero. There are two types of managers, high ability, referred to as good (g), with probability (w.p.) $1-\theta$ and low ability, referred to as bad (b), w.p. θ. Managers are employed by firms that have made a previous investment (before $t = 0$) that is generating a cash flow stream at $t = 1$ and $t = 2$, after which the firm will be liquidated. At $t = 1$ firms have cash flows (CFs) of high (H) w.p. δ_i^H, medium (M) w.p. δ_i^M, or low (L) w.p. δ_i^L where $H >> M > L > 0$ and $i \in \{g,b\}$. I assume good managers are more likely to have higher cash flows, equally likely to have medium cash flows, and less likely to have lower cash flows compared to bad managers. That is, $\delta_g^H > \delta_b^H$, $\delta_g^M = \delta_b^M$, and $\delta_g^L < \delta_b^L$. Further, I assume that realized CFs are observable by the market at the end of the period and managerial type is unknown but updatable using Bayes rule.[4] At $t = 2$ firms have CFs that will depend on their realized CF_1 as follows: (1) a firm with $CF_1 = H$ will have a CF_2 that is very high (HH), high (H), or medium (M) w.p. δ_i^{HH}, δ_i^H, and δ_i^M, respectively; (2) a firm with $CF_1 = M$ will have a CF_2 that is high (H), medium (M), or low (L) w.p. δ_i^H, δ_i^M, and δ_i^L, respectively; and (3) a firm with $CF_1 = L$ will have a CF_2 that is medium (M), low (L), or very low (LL) w.p. δ_i^M, δ_i^L, and δ_i^{LL}, respectively.[5] The CFs at $t = 2$ for good and bad managers are dependent upon the CFs at $t = 1$. Therefore, one can view the CFs over time as revealing some information about changes in permanent earnings.

All firms, regardless of the manager's type, are assumed to have just paid a dividend of $d_0 = M$ at $t = 0$.[6] If CFs are not fully paid out as a dividend, the surplus CFs $(CF_t - d_t)$ is carried forward at a per dollar cost, cs_i where $0 \leq cs_i < 1$ and $i \in \{g,b\}$. As indicated by Jensen's (1986) free-cash-flow hypothesis, excess cash introduces a moral hazard problem as managers may be more likely to frivolously spend this money. Further, Stulz (1990) shows that managers may wish to invest in negative NPV projects since their perquisites will increase. Therefore, one can view good managers as those who have better control within the firm to not waste money. Thus, good managers are assumed to have lower cost to carrying cash forward than bad managers, specifically $cs_b > cs_g = 0$.[7] If CFs are less than the dividend promised, the manager must borrow the shortage at a cost, $\overline{R}_i \in \{\underline{R}_i, \overline{R}_i\}$ where $0 < \underline{R}_i < \overline{R}_i < 1$ and $i \in \{g,b\}$. I assume that good managers are able to borrow at lower costs than bad managers.[8] Further, I assume the marginal borrowing costs increase with the amount the manager must borrow; that is, borrowing costs exhibit the following ordering: $\underline{R}_g < \underline{R}_b \leq \overline{R}_g < \overline{R}_b$.

At $t = 0$ managers know their type and CF_1 but not CF_2. At $t = 1$ managers receive a perfect signal regarding CF_2. The manager would like to signal the firm's future cash flows by changing the dividend from d_0.[9] This signal would increase the firm's market price, which in turn increases the manager's utility. However, the manager also has a reputational concern. He would like the market to believe he is of a good type since this raises his reservation wage. To capture this an exogenous bonus, B, is added to the manager's utility if the manager is known to be good with probability one. Thus, there are two things the manager would like to signal, the firm's future CFs and his ability. However, dividends provide the only signaling mechanism. Therefore, a signal jamming problem results.

If both good and bad managers increase dividends when $CF_1 = H$, then the firm's future CFs are revealed but the good manager is pooled with the bad manager. Conversely, if $CF_1 = L$ and both managers cut dividends, then good and bad managers are again pooled. However, since a good manager has lower financing and carrying costs, a good manager with $CF_1 = H$ (L) will announce at $t = 0$ a dividend less than (greater than) CF_1. Given that the market observes CF_1, any manager that either obtains outside financing or incurs a carrying cost will be viewed as a good manager. Thus, the good manager is able to separate from the bad manager by not changing his promised dividend to equal CF_1, even though he knows CF_1 perfectly. At $t = 2$, all firms will pay a liquidating dividend. If a manager signals a higher firm value than is true at $t = 1$, the manager loses the reputational bonus and also incurs a personal bankruptcy cost, π.

A time line of the model is depicted in Fig. 1.

Each manager is assumed to maximize his expected utility subject to the dividend paid; i.e.

$$\max_{\text{s.t.}\, d_t} \; E_t[U_i] = \alpha \times P_t^m(d_t) + \beta \times P_{t+1}^T + \lambda_{t+1} \times B - \pi$$

where d_t is the dividend paid at time t, $\alpha, \beta \in [0,1]$ are constants, $P_t^m(d_t)$ is the market's value of the firm at time t, P_{t+1}^T is the true value of the firm at time t+1, λ_{t+1} is the updated probability a manager is good, and π is the personal bankruptcy costs of the manager.[10,11]

The value of the firm depends on the dividend the manager pays. The *true* $t = 2$ cum-dividend value of a firm given CF_1 and d_1 is

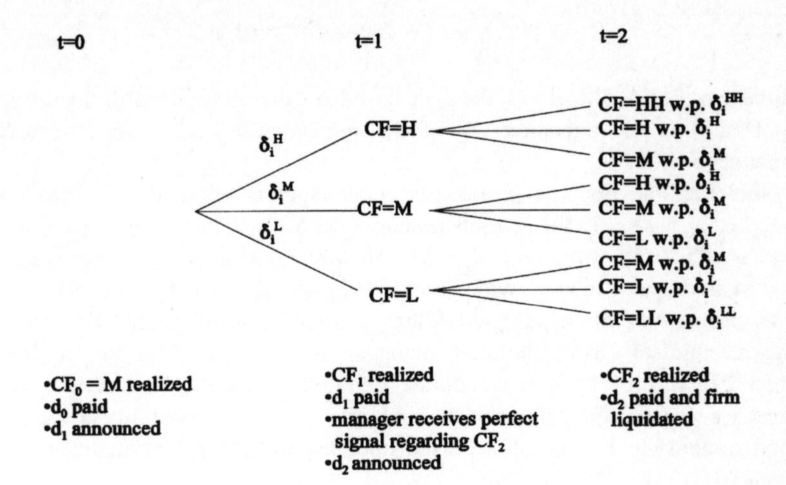

Fig. 1. Time Line.

$$P_2^T = \begin{cases} CF_1 - d_1 + CF_2 & \text{if } CF_1 - d_1 = 0 \\ (CF_1 - d_1)\times(1 - cs_i) + CF_2 & \text{if } CF_1 - d_1 > 0 \\ (CF_1 - d_1)\times(1 + R_i) + CF_2 & \text{if } CF_1 - d_1 < 0 \end{cases} \quad (1)$$

and the *true* $t = 1$ cum-dividend value of a firm is

$$P_1^T = \begin{cases} CF_1 - E_i[CF_2^j] & \text{if } CF_1 - d_1 = 0 \\ (CF_1 - d_1)\times(1 - cs_i) + d_1 + E_i[CF_2^j] & \text{if } CF_1 - d_1 > 0 \\ (CF_1 - d_1)\times(1 + R_i) + d_1 + E_i[CF_2^j] & \text{if } CF_1 - d_1 < 0 \end{cases} \quad (2)$$

where $i \in \{g,b\}$, $j \in \{H,M,L\}$, and $E_i[CF_2^j]$ is the expected CF_2 when $CF_1 = j$. The market perceived value of the firm at $t = 0$ and $t = 1$ will depend upon the dividend announced in those period, i.e. d_1 and d_2. Define $P_t^m(d_t)$ to be the market's valuation of the firm at time t given d_t. At $t = 1$, then

$$P_1^m(d_2) \begin{cases} CF_1 - d_1 + CF_2(d_2) & \text{if } CF_1 - d_1 = 0 \\ (CF_1 - d_1)\times(1 - cs_i) + CF_2(d_2) & \text{if } CF_1 - d_1 > 0 \\ (CF_1 - d_1)\times(1 + R_i) + CF_2(d_2) & \text{if } CF_1 - d_1 < 0 \end{cases} \quad (3)$$

where $CF_2(d_2)$ is the market's believed CF_2 based on the announced dividend d_2, and at $t = 0$

$$P_o^m(d_1) = CF_1(d_1) + E[CF_2(d_1)] \quad (4)$$

where $i \in \{g,b\}$, $CF_1(d_1)$ is the CF_1 based on the announced dividend d_1, and $E[CF_2(d_1)]$ is the expected CF of the manager at $t = 2$ given the dividend announced at $t = 0$.

Note that the updated probability a manager is good at $t = 1$ and $t = 2$ is one, i.e. $\lambda_1 = \lambda_2 = 1$, for a good manager who has $CF_1 = H$ and pays $d_1 = d_1^*$ or has $CF_1 = L$ and pays $d_1 = M$. However, a good manager that pays $d_1 = M$ and has $CF_1 = M$ will have $\lambda_1 = \delta g \times \theta / [\delta_g^M \times \theta + \delta_b^M \times (1-\theta)] = \theta$, his prior probability of being good. Thus, a good manager with $CF_1 = M$ cannot separate himself form the bad manager with $CF_1 = M$ at $t = 1$. However, when $CF_2 = H$ or L is realized, the updated probability a manager is good increases (decreases). Yet, if $CF_2 = M$, the updated probability a manager is good remains θ. Let $\lambda_2(CF_2)$ be the updated probability the manager is good given $CF_1 = M$.[12]

The proposed equilibrium dividend policy, that will be shown to hold later, has bad managers announcing $d_1 = CF_1$, good managers with $CF_1 = H$ announcing $d_1^* < H$, and good managers with $CF_1 = M$ or $CF_1 = L$ announcing $d_1 = d_0 = M$. Thus, for firms announcing $d_1 = M$, $P_0^m(d_1)$ is a weighted average of good and bad managers' CFs and dividends. That is, $P_0^m(d_1 = M) = CF_1(d_1 = M) + E[CF_2(d_1 = M)]$ where

$$CF_1(d_1 = M) = (\mu_g^M + \mu_b^M) \times [M - d_1 + d_1] + \mu_g^L \times [(L - d_1) \times (1 + R_g) + d_1],$$

$$E[CF_2(d_1 = M)] = \mu_g^M \times E_g[CF_2^M(d_1)] + \mu_b^M \times E_b[CF_2^M(d_1)] + \mu_g^L \times E_g[CF_2^L(d_1)],$$

μ_i^j is the probability of $d_0 = d_1 = M$ for a manager of type i with $CF_1 = j$, and $E_i[CF_2^j(d_1)]$ is the expected CF_2 plus or minus any net surplus or borrowing for a manager of type i with $CF_1 = j$, where $i \in \{g,b\}$ and $j \in \{H,M,L\}$.[13]

I begin analyzing the model's second period. The bad manager will announce $d_2 = CF_2$ since he incurred no carrying or borrowing costs from period one. The liquidating dividends for the good managers are based on the proposed equilibrium strategies at $t = 0$ and CF_2. Good managers pay $d_2 > CF_2$ if $CF_1 = H$, $d_2 = CF_2$ if $CF_1 = M$, and $d_2 < CF_2$ if $CF_1 = L$. Since good managers with $CF_1 = H$ pay $d_1 < H$, excess cash exists that must be paid at $t = 2$, making $d_2 > CF_2$. Further, good managers with $CF_1 = L$ pay $d_1 = M$, incurring borrowing costs that must be repaid at $t = 2$, thus, $d_2 < CF_2$.

Proposition 1: If $\pi \geq -\alpha \times (M - HH)$, then the following is a separating equilibrium:

(i) a pure strategy for the good managers to announce $d_2 = CF_2+(H-d_1)$ if $CF_1 = H$, $d_2 = CF_2$ if $CF_1 = M$, or $d_2 = CF_2-(M-L)(1+\underline{R}_g)$ if $CF_1 = L$, and a pure strategy for the bad managers to announce $d_2 = CF_2$;
(ii) a date-1 signal price of $P_1^m(d_2)$ given by equation (3); and
(iii) a date-2 post cash flow realization price of P_2^T given by equation (1).

The market will revise its beliefs about the manager's type and the firm's future CFs after observing the dividend signal. If the dividend signal is one of the equilibrium dividend levels or equal to CF_2, then the market uses Bayes rule to value the firm and assess the manager's type. If the manager chooses an out-of-equilibrium dividend level, then the market believes with probability one that he is bad.

The intuition behind Proposition 1 is as follows. Since all CFs are known at the end of period one and uncertainty is resolved for managers with high CFs or low CFs, bad or good managers cannot mimic the other type but can mimic a similar type manager with a higher CF_2 or the same CF_1. For example, a good manager with $CF_1 = H$ and $CF_2 = M$ might be tempted to mimic another good manager with $CF_1 = H$ and $CF_2 = HH$. The constraint that binds is the one that ensures a bad manager with $CF_2 = M$ does not mimic another bad manager with $CF_2 = HH$. Intuitively this follows since $H >> M$ and a bad manager only incurs a personal cost for lying of π, while a good manager would incur π and lose his bonus. Therefore, if bad managers do not mimic, neither will good managers. If the personal cost for lying, π, is greater than the added utility from falsely signaling with a higher dividend, $\alpha(HH-M)$, this equilibrium will hold.

Now rolling back to $t = 0$, I show that managers do indeed follow the proposed dividend policy at $t = 0$. Since all information is revealed at the end of period 1, the actions taken in period 2 will not impact the period 1 decision. Therefore, I can solve the model as if it is the union of two separate one-period dividend signaling games.

Lemma 1: The two period incentive compatibility constraints will hold when solving the model as two separate one-period signaling games.

The intuition underlying Lemma 1 is that since all asymmetric information is resolved at $t = 1$ for $CF_1 = L$ or H, there can be no mimicking when d_2 is announced, even if there is mimicry at $t = 0$. Thus, the two-period incentive compatibly constraints are satisfied trivially. Further, when $CF_1 = M$, both good and bad managers are observationally equivalent, so there can be no mimicry.

The proposed equilibrium dividend level at $t = 1$ for bad managers is $d_1 = CF_1$. For good managers if $CF_1 = M$ or L, then $d_1 = M$. However, if $CF_1 = H$, $d_1 = d_1^*$. d_1^* is the dividend that achieves incentive compatibility between bad and good managers. That is, a bad manager with $CF_1 = H$ must be indifferent between announcing $d_1 = H$ and $d_1 = d_1^*$. The d_1^* that achieves incentive compatibility satisfies the following:

$$E_0[U(d_1 = H)] = \alpha\times[H+\{\delta_b^{HH}\times HH+\delta_b^{H}\times H+\delta_b^{M}\times M\}]+$$
$$\beta\times[H+\{\delta_b^{HH}\times HH+\delta_b^{H}\times H+\delta_b^{M}\times M\}]$$
$$\geq$$
$$E_0[U(d_1 = d_1^*)] = \alpha\times[H-d_1^*)\times(1-cs_g)+ d_1^*+\{\delta_g^{HH}\times HH+\delta_g^{H}\times H+\delta_g^{M}\times M\}]$$
$$+ \beta\times[H-d_1^*)\times(1-cs_b)+d_1+\{\delta_b^{HH}\times HH+\delta_b^{H}\times H+\delta_b^{M}\times M\}]$$

where $E_0[U(\bullet)]$ is the expected utility of a bad manager signaling with d_1. This equation simplifies to

$$d_1^* = \frac{\alpha\times[(\delta_b^{HH} - \delta_g^{HH})\times HH+(\delta_b^{H} - \delta_g^{H})\times H+(\delta_b^{M} - \delta_g^{M})\times M]}{[\alpha\times cs_g+\beta\times cs_b]}+H. \quad (5)$$

Note that $d_1^* < H$ if the first the first term of the right hand side of the equation (5) is less than 0. This will be true since by assumption $\delta_b^{HH} < \delta_g^{HH}$, $\delta_b^{H} = \delta_g^{H}$, and $\delta_b^{M} > \delta_g^{M}$ and $HH > H >> M$.[14]

I can now characterize the Nash equilibrium at $t = 0$.

Proposition 2: There exist values of the exogenous parameters and prior beliefs such that the following is a separating equilibrium in which it is:

(i) a pure strategy for the good managers to announce $d_1 = M$ if $CF_1 = L$ or M or $d_1 = d_1^*$ if $CF_1 = H$, and a pure strategy for the bad managers to announce $d_1 = CF_1$;
(ii) a date-0 signal price of $P_0^m(d_1)$ given by equation (4); and
(iii) a date-1 post-cash flow-realization price of P_1^T given by equation (2).

The market will revise its beliefs about the manager's type and the firm's future CFs after observing the dividend signal. If the dividend signal is one of the equilibrium dividend levels or equal to CF_2s, then the market uses Bayes rule to value the firm and assess the manager's type. If the manager chooses an out-of-equilibrium dividend level, then the market believes with probability one that he is bad.

The intuition underlying Proposition 2 is as follows. Since managerial type is unknown when the dividend is announced at $t = 0$, the constraint that binds is the one that ensures a bad manager with high CFs does not mimic a good manager with high CFs. As shown in equation (5), this constraints holds tightly.

However, it must also hold that bad managers with low CFs do not mimic good managers with medium CFs and good managers with medium CFs do not mimic good managers with high CFs. These constraints depend on the cost of borrowing for good and bad managers. If the cost of borrowing is significant, then the benefit of falsely signaling and obtaining a greater value at $t = 1$ is overwhelmed by the losses faced at $t = 2$.

Further, the model shows.

Proposition 3: Good managers smooth dividends more than bad managers.

Since good manager with higher CFs pay $d_1^* < H$ and bad managers pay $d_1 = H$, good managers smooth dividends more than bad managers and relative to earnings. When CFs are low, good managers announce $d_1 = d_0 = M$ and bad managers announce $d_1 = L$. Thus, by keeping dividends inflated when earnings are low, good managers smooth dividends relative to earnings and also more than bad managers. Thus, good managers smooth dividends in order to signal the firm's future cash flows and also their ability.

III. EMPIRICAL HYPOTHESES

Several new empirical implications beyond Lintner (1956) emerge from the above model.

Hypothesis A: Dividend increases are good news and dividend decreases are bad news across all firm types.

This prediction is consistent with the static dividend signaling models and has been tested many times.[15]

Hypothesis B: Firms that smooth more will have larger price increases to an unexpected dividend increase and a smaller price decline to an unexpected dividend decrease.

As shown by the model, firms that smooth more have better managers and higher future cash flows, and thus, greater (smaller) price appreciation (depreciation) to an unexpected dividend increase (decrease). This implication can be empirically verified by examining the associated price reaction for unexpected dividend increases and decreases conditional upon the amount of dividend smoothing prior to the unexpected change. Though this result has not been empirically tested in previous studies, this result is not a prediction from this model alone. In fact, all dividend smoothing models implicitly predict that firms that smooth more will have larger price increases to unexpected dividend increases. Therefore, empirical support of this prediction only lends credence to this model but does not prove this model explains dividend smoothing.

Hypothesis C: Firms that smooth dividends more have higher values.

As indicated in Proposition 3, for both high and low cash flows, good managers will smooth their firms' dividends because it will lead to higher firm market values and higher reputational values. This empirical prediction is novel in that it ties dividend smoothing to firm value. Previous theoretical and empirical work shows that firms do smooth dividends, but do not investigate the future performance of firms that have smoothed dividend more in the past. The implication that greater dividend smoothing leads to higher future values can easily be tested; however, the implication that greater dividend smoothing leads to higher reputation values is not easily verifiable since managerial reputation is not as widely reported or accepted. Thus, I test whether firms that have smoothed more prior to a dividend change have higher future values after the dividend change.

Hypothesis D: Firms with good managers will smooth more than bad managers.

The theory predicts that firms that have good managers should smooth dividends more than bad managers. This prediction can be tested using CEO tenure as a proxy for managerial type. The assumption made is that the longer the CEO has been tenured, the more likely it is that the CEO is "good," as it becomes harder and harder for a manager to continually fool the market.

Hypothesis E: Firms with smaller dividend increases have larger price increases.

This empirical prediction is perhaps the most novel of the model. All else equal, high-value firms with good managers will smooth dividends more. Thus, their smaller dividend increases should garner greater price reactions.

IV. EMPIRICAL TESTS

To test the model's predictions, I collect dividend changes from the Center for Research in Security Prices (CRSP) monthly master file over the period 1989 to 1993. For a firm having a dividend increase or decrease to be included in the sample, the following criteria must be satisfied:

(i) The dividend change must be associated with a regular quarterly U.S. cash dividend per share.

(ii) The announcement does not represent a dividend initiation or the first dividend since a dividend omission.
(iii) The firm has a *Value Line* forecasted yearly dividend for each year from 1982 to the announcement year.
(iv) The firm does not have an earnings announcement or other information publicized within a three-day window of the announced dividend change.[16]
(v) A stock split or stock dividend does not occur in the month before or the month after the dividend change announcement is made.
(vi) Daily return data for the 230 trading days surrounding the dividend change announcement are available from the CRSP daily return file.
(vii) The firm must be listed on the Compustat quarterly files for at least three years before the unexpected dividend change.
(viii) CEO tenure data is available from *Forbes* for the year of the unexpected dividend change.[17]
(ix) The firm is not a utility company, defined by the first two-digits of the SIC code of 49.

These restrictions result in a sample of 746 unexpected dividend increases and 172 unexpected dividend decreases.

The unexpected dividend change is calculated as the difference between the quarterly dividend paid and the expected quarterly dividend. The *Value Line* annual dividend prediction is used as the proxy for the expected dividend. The quarterly expected dividend is assumed to be the annual prediction divided by four. If the unexpected dividend change is greater than zero, the expected dividend for next quarter is the new dividend paid. For example, if *Value Line* predicts a yearly dividend of $1.00, the expected quarterly dividend is $0.25. If the actual quarterly dividend paid in the first quarter is $0.20, the unexpected dividend change equals −$0.05, and the expected quarterly dividend remains at $0.25. However, if the actual quarterly dividend is $0.30, the unexpected dividend change is $0.05, and the expected dividend change becomes $0.30.[18]

First, I test Hypothesis A: that unexpected dividend increases (decreases) result in positive (negative) stock price reactions. Using the market-model methodology outlined in Brown and Warner (1980, 1985), the ordinary-least-squares coefficients of the market model regression are estimated over the period $t = -200$ to $t = -30$. The daily abnormal stock return is calculated for each firm i for the announcement day, $t = 0$, and the announcement day average abnormal return (AR) is computed across all firms. The AR on the announcement day is 0.40% for unexpected dividend increases and −1.39% for unexpected dividend decreases, both of which are highly significant.

Second, I analyze Hypothesis B: firms that have smoothed dividends more in the past have greater (smaller) price reactions to unexpected dividend increases (decreases). The smoothing variable is measured as the ratio of the standard deviation in quarterly dividends to the standard deviation in quarterly earnings for three years prior to the unexpected dividend change.[19] Note, a higher smooth variable implies less dividend smoothing. Since the prediction indicates that, regardless of whether there is an unexpected dividend increase or decrease, firms that have smoothed more in the past will have higher price reactions, I estimate the following for all unexpected dividend changes using weighted least squares[20]:

$$CAR_{t=-1,0} = \beta_1 LSIZE + \beta_2 SMOOTH + \beta_3 DIVYLD + \beta_4 VOLUME + \beta_5 MKTBK + \beta_{i+4} Ii + \epsilon \qquad (6)$$

where $CAR_{t=-1,0}$ is the cumulative abnormal return over the two-day period, $t = -1$ to $t = 0$, LSIZE is the log of the market capitalization of the firm as of one month prior to the unexpected dividend change, SMOOTH is the measure of dividend smoothing, DIVYLD is the firm's dividend yield for the year prior to the announcement, VOLUME is the log of the average daily trading volume, MKTBK is the firm's market-to-book ratio as of one month prior to the unexpected dividend change, $Ii (i \in \{2,9\})$ are dummy variables for the eight industry classifications, and ϵ is the ordinary least squares error.

The log of the firm size is included to control for the fact that smaller firms generally have higher returns. The dividend yield is included to capture the clientele effect documented by Bajaj and Vijh (1990) and Denis, Denis and Sarin (1994). Volume is included since Eberhart and Damodaran (1997) find it a significant determinant of abnormal returns surrounding earnings announcements. The market-to-book ratio is included as some growth firms (those with large market-to-book ratios) are less likely to pay dividends while firms with fewer growth opportunities are more likely to pay dividends and smooth.[21] I include dummy variables for eight different industries classifications as some industries may have more dividend smoothing than others. Table 1 summarizes the variables used in this study and defines the industries classifications.

Table 2 contains the results from the regression in equation (6).[22] There exists a negative and significant relation between dividend smoothing and CARs. That is, firms that have smoothed more have larger price appreciations to unexpected dividend changes. Volume has an insignificant impact on CARs, while the past dividend yield has a negative and significant impact on CARs. This seems to support the idea that there is a dividend clientele effect. The size of the firm has a significantly positive impact on CARs which differs from standard

Table 1. Summary Statistics for NYSE-, AMEX-, and NASDAQ-listed Firms with Unexpected Dividend Changes from 1989 to 1993.

Variable	N	Mean	Standard Deviation	Minimum	Maximum
Tenure as CEO	572	8.0455	7.5371	0	40
Tenure with firm	572	24.7325	12.1631	0	51
Firm Age	572	57.1521	5.7159	39	80
Firm Size	918	4,426,025,840	8678748.10	32,295,880	83,279,336,000
Market-to-Book	918	2.6766	2.5161	0.5423	33.3330
Dividend Yield	918	0.0314	0.03152	0.0025	0.3830
Smoothing Proxy		0.3224	0.2714	0.0000	2.4495
$CAR_{t=-1,0}$	918	–0.0006	0.0363	−0.2298	0.1562
Volume	918	1485118.32	2443268.48	3000.00	21486901.00
Quarterly Dividend Change	918	0.0043	0.1261	−1.4400	1.2600
Unexpected Dividend Change	918	-0.0039	0.1205	−0.6500	1.5500

Industry Number:	Industry Classifications:	First two digits of SIC Code:	Number of Firms in Each Industry:
1	Agriculture, Forestry, and Fisheries	01, 02, 07, 08, 09	0
2	Minerals	10, 12, 13, 14	40
3	Construction	15, 16, 17	14
4	Manufacturing	20–39	491
5	Transportation and Communications	41–49	56
6	Wholesale Trade	50, 51	26
7	Retail Trade	52–59	60
8	Finance, Insurance, and Real Estate	60–67	181
9	Services	70, 72, 73, 75, 76, 78–84, 86–89	50
10	Public Administration	91–97	0

empirical findings. However, since both positive and negative unexpected dividend changes are grouped together and negative changes would have a positive relation with size, this result is not surprising.[23] The coefficients on the industry dummies are all negative buy only significant for firms in transportation and communications, retail trade, and financials.

Third, Hypothesis C is tested: firms that smoothed more in the past have higher values subsequent to an unexpected dividend change. I estimate value using two proxies: (1) the market-to-book ratio and (2) the market value of common stock plus the market value of preferred stock[24] plus the book value

Table 2. Regression of the Average Standardized CAR Over the Two-day Period, t = −1 to t = 0, on the Firm's Size, the Firm's Measure of Smoothing, the Firm's Dividend Yield for the Year Prior to the Announcement, the Firm's Average Trading Volume, and Dummy Variable for the Eight Industry Classifications for NYSE-, AMEX-, and NASDAQ-listed Firms with Unexpected Dividend Changes from 1989 to 1993. T-statistic Listed in Parenthesis Below Coefficient Estimate.

LSIZE	0.003358 (2.596)***
SMOOTH	–0.014838 (–2.906)***
DIVYLD	–0.090678 (-1.817)*
VOLUME	–0.001804 (–1.472)
MKTBK	0.000637 (1.317)
I2	–0.014857 (–1.177)
I3	–0.006619 (–0.428)
I4	–0.017603 (–1.628)
I5	–0.028554 (–2.430)**
I6	–0.009849 (–0.807)
I7	–0.022800 (–1.957)**
I8	–0.020860 (–1.879)*
I9	–0.010580 (–0.932)
N	918
Adjusted R^2	5.96%
F-Statistic	4.410***

*,**, and *** indicate significance at the 10%, 5%, and 1% level, respectively.

of liabilities minus any deferred taxes and investment tax credits. The change in future value is the percentage change in the proxy for the year after the unexpected dividend change. I regress the percentage change in value on the measure of smoothing, the log of the firm's size, and the industry dummy variables; i.e.

$$\%\Delta\text{VALUE} = \beta_1\text{SMOOTH}+\beta_2\text{LSIZE}+\beta_{i+1}\text{Ii}+\epsilon \tag{7}$$

where %ΔVALUE is the percentage change in the proxy for market value and all other variables are defined as before.

Column 2 of Table 3 shows that when the change in the market-to-book ratio is used as a proxy for value, smoothing does not significantly impact the future value. However, the firm size has negative and significant impact on the change in value. That is, the larger the firm size at the dividend announcement, the lower the future change in value. All the industries have a positive and significant impact on value except for construction, wholesale trade, and retail trade. However, Column 3 indicates that when equity plus liabilities are used to proxy for value, there is a negative and highly significant relation between smoothing and future value. This supports the model's predictions that firms which smooth more prior to an unexpected dividend change have higher values in the future. Firm size has no impact on future value, while firms in manufacturing, wholesale trade, financials and services have a positive and significant impact on future value. One reason for the insignificant results for the market-to-book ratio may be that the market-to-book is a noisy proxy for value. Market-to-book ratios are often used as proxies for the firm's growth opportunities, value, reputation, etc.

Fourth, I test Hypothesis D: firms with good managers smooth more than firms with bad managers. The CEO's tenure with the firm and the CEO's tenure as CEO are both used as proxies for managerial type. I regress the CEO's tenure on the measure of smoothing, the age of the firm, and industry dummy variables. The firm's age is included since older firms have more stable dividend payments.[25] I estimate the following:

$$\text{SMOOTH} = \beta_1\text{CEOTENURE}+\beta_2\text{FIRMAGE}+\beta_{i+1}\text{Ii}+\epsilon \tag{8}$$

where CEOTENURE is either the number of years the CEO has worked for the firm or the number of years the CEO has been CEO, FIRMAGE is the number of years the firm has existed, and all other variables are defined as before.

As indicated in Table 4 the coefficient on CEOTENURE is negative and highly significant for both proxies. Therefore, the longer a CEO has been either CEO or with the firm, the more likely he is to smooth. In Column 2, when

Table 3. Regression of the Percentage Change in the Firm's Value on the Measure of Dividend Smoothing, the Firm's Log Size, and Dummy Variables for the Eight Industry Classifications for NYSE-, AMEX-, and NASDAQ-listed Firms with Unexpected Dividend Changes from 1989 to 1993 . T-statistic Listed in Parenthesis Below Coefficient Estimate.

Regression	%ΔVALUE = percentage change in the market-to-book ratio	%ΔVALUE = percentage change in the market value of equity plus debt
LSIZE	–0.018666 (–2.086)**	–0.005800 (–0.634)
SMOOTH	0.002165 (0.039)	–0.145581 (–2.581)***
I2	0.374801 (2.514)**	0.227945 (1.495)
I3	0.251336 (1.503)	0.174256 (1.019)
I4	0.322309 (2.492)**	0.218116 (1.649)*
I5	0.299687 (2.127)**	0.199354 (1.384)
I6	0.089820 (0.605)	0.263693 (1.736)*
I7	0.186188 (1.323)	0.183341 (1.274)
I8	0.259429 (1.988)**	0.291100 (2.181)**
I9	0.279238 (2.046)**	0.240637 (1.724)*
N	918	918
Adjusted R^2	2.37%	6.47%
F-Statistic	2.206**	6.283***

*,**, and *** indicate significance at the 10%, 5%, and 1% level, respectively.

CEO tenure is measured as the number of years the CEO has been CEO, the older the firm, the less likely the manager is to smooth dividends. Construction, the only industry with a significant impact, has a negative effect on smoothing. However, as reported in Column 3, when CEO tenure is the CEO's tenure with the firm, the firm's age has no impact on smoothing. Manufacturing and

Table 4. Regression of the Measure of Dividend Smoothing on CEO Tenure, Firm Age, and Dummy Variables for the Eight Industry Classifications for NYSE, AMEX-, and NASDAQ-listed Firms with Unexpected Dividend Changes from 1989 to 1993. T-statistic Listed in Parenthesis Below Coefficient Estimate.

Regression	CEOTENURE = number of years the manager has been CEO of the firm	CEOTENURE = number of years the CEO has worked for the firm
CEOTENURE	–0.003568 (–3.003)***	–0.003702 (–5.229)***
FIRMAGE	0.006187 (4.006)***	0.001732 (1.148)
I2	–0.080927 (–0.761)	0.079596 (0.778)
I3	–0.295936 (–2.506)**	–0.124612 (–1.098)
I4	0.006394 (0.073)	0.143572 (1.722)*
I5	0.146525 (1.615)	0.296948 (3.416)***
I6	0.089820 (0.605)	0.040844 (0.434)
I7	–0.066550 (–0.686)	0.010791 (0.129)
I8	–0.090730 (–1.055)	0.042323 (0.510)
I9	–0.091748 (1.014)	0.046430 (0.525)
N	572	572
Adjusted R^2	72.23%	73.09%
F-Statistic	149.716***	156.323***

*,**,and *** indicate significance at the 10%, 5%, and 1% level, respectively.

transportation and communications both have positive and significant impacts on smoothing, while all other industries have no impact.

Finally, I verify Hypothesis E: firms with smaller dividend increases have larger price increases. The sample of unexpected dividend increases are divided into those firms that smoothed more and those that smoothed less in the past.

This division is based on the midpoint of the smoothing variable for the sample of firms with unexpected dividend increases. Then, dummy variables are created that quartile the firms based on the size of the unexpected dividend increase. For the two samples, firms that smooth less and those that smooth more, I estimate the following using weighted least squares:

$$CAR_{t=-1,0} = \beta_1 Q1 + \beta_2 Q2 + \beta_3 Q3 + \beta_4 Q4 + \beta_5 LSIZE + \beta_6 DIVYLD \quad (9)$$

$$+ \beta_7 VOLUME + \beta_8 MKTBK + \beta_{i+7} Ii + \epsilon$$

where Q1 is 1 if the unexpected dividend change is in the lowest 25% and 0 otherwise, Q2 is 1 if the unexpected dividend change is in the next 25% and 0 otherwise, Q3 is 1 if the unexpected dividend change is in the next 25% and 0 otherwise, Q4 is 1 if the unexpected dividend change is in the highest 25% and 0 otherwise, and all other variables are as previously defined.

Column 2 of Table 5 shows that for firms that have smoothed more in the past, Q1, Q2, and Q3 all show a positive and significant relation between smoothing and CARs. However, Q4 has no significant impact on CARs. Further, the coefficients for Q1, Q2, Q3, and Q4 are monotonically decreasing, and an F-test of whether the coefficients are significantly different than each other is significant at the 5% level (F-value = 1.97). Firm size, volume, and market-to-book ratio have no significant impact on CARs, while the dividend yield has a positive and significant impact. Minerals and financials have a negative and significant impact on CARs. The results in Column 3 indicate that the same relation does not hold for firms that have smoothed less in the past. The coefficients for Q1, Q2, Q3, and Q4 are all insignificant. Further, firm size, dividend yield, volume, and market- to-book ratio are all insignificant. All industries have a negative impact on CARs except for construction and wholesale trade. Thus, the evidence supports the prediction that high value firms with good managers that have smoothed dividends should have greater positive price reactions to smaller unexpected dividend increases.

V. CONCLUSION

This paper examines how the desire of the manager to develop a good reputation can influence the dividend policy of the firm. The model in this paper derives an equilibrium in which good managers, with private knowledge about the firm's future cash flows and their ability, smooth dividends to signal to the market these two characteristics. Good managers with high cash flows will pay lower dividends and carry excess cash into the future, while bad managers pay out all cash as dividends to avoid the high carrying costs. When future cash flows

Table 5. Regression of the average standardized CAR over the two–day period, t = –1 to t = 0, on indicator variables for the quartiles of the unexpected dividend change, the firm's size, the firm's measure of smoothing, the firm's dividend yield for the year prior to the announcement, the log of the firm's average trading volume, the firm's market–to–book ratio, and dummy variables for the industry classifications for NYSE-, AMEX-, and NASDAQ-listed firms with unexpected dividend increases from 1989 to 1993 Firms have been divided in half based on the smoothing prior to the unexpected dividend increase. T-statistic listed in parenthesis below coefficient estimate.

Regression	more past smoothing (lower SMOOTH variable)	less past smoothing (higher SMOOTH variable)
Q1	0.029157	0.004077
	(1.876)*	(0.289)
Q2	0.028547	0.000668
	(1.766)*	(0.048)
Q3	0.028017	0.005577
	(1.758)*	(0.391)
Q4	0.026108	0.011608
	(1.544)	(0.780)
LSIZE	–0.002145	0.000226
	(–1.258)	(0.143)
DIVYLD	0.328297	0.114535
	(2.549)**	(1.270)
VOLUME	0.000535	0.000543
	(0.356)	(0.336)
MKTBK	0.000567	0.000345
	(0.243)	(0.628)
I2	–0.017035	–0.027043
	(–1.673)*	(–2.585)***
I3	0.009600	–0.013078
	(0.757)	(–0.637)
I4	–0.006700	–0.016765
	(–1.200)	(–2.618)***
I5	–0.008913	–0.023488
	(–1.095)	(–3.000)***
I6	0.002475	–0.006627
	(0.289)	(–0.591)
I7	–0.010704	–0.018547
	(–1.468)	(–2.401)**
I8	–0.012002	–0.014677
	(–2.008)**	(–2.137)**
N	373	373
Adjusted R^2	6.86%	10.76%
F-Statistic	1.888**	3.093***

*,**,and *** indicate significance at the 10%, 5%, and 1% level, respectively.

are low, good managers will borrow at a more favorable rate than bad managers and keep dividends higher. Therefore, good managers smooth dividends more than bad managers and also relative to earnings.

The model also generates several testable empirical predictions beyond Lintner (1956). First, firms that smoothed dividends more in the past will have larger price appreciations to unexpected dividend increases and smaller price declines to unexpected dividend decreases. Second, firms that have smoothed dividends more in the past will have higher future values. Third, good managers will smooth dividends more than bad managers. Fourth, firms with smaller dividend increases have larger price increases. The empirical evidence suggests that these predictions do hold.

APPENDIX

Proof of Proposition 1: First I show it is incentive compatible for the good managers and bad managers to follow the equilibrium strategies stated in Proposition 1. Note that the subscript in the utility indicates the CF_2. Since all information asymmetry is resolved regarding the manager's type when CF_1 are realized and d_1 is paid, I only need to ensure that good managers do not mimic other good managers with higher CFs and bad managers do not mimic other bad managers with higher CFs.

WHEN $CF_1 = H$

Note for all the following IC constraints I have assumed if the manager lies he loses his bonus, B, in addition to paying π. Further, I will only check the case where the lowest cash flow would mimic the highest cash flow (in this case M and HH) since if this one binds, I know that the rest will also bind. That is, M will not mimic H and H will not mimic HH).

- good manager with $CF_2 = M$ does not mimic good manager with $CF_2 = HH$

$$E_1[U_{M|M}(CF_1 = H, CF_2 = M)] = \alpha\times[M+\epsilon\times(1-cs_g)]+\beta\times[M+\epsilon\times(1-cs_g)]+1\times B\geq$$

$$E_1[U_{HH|M}(CF_1 = H, CF_2 = HH)] = \alpha\times[HH+\epsilon\times(1-cs_g)]$$

$$+\beta\times[M+\epsilon\times(1-cs_g)]+0\times B-\pi$$

For the above to bind then,

$$\pi\geq -\alpha \times (M-HH) - 1\times B.$$

- bad manager with $CF_2 = M$ does not mimic bad manager with $CF_2 = HH$

$$E_1[U_{M|M}(CF_1 = H,\ CF_2 = M)] = \alpha\times[M]+\beta\times[M]+0\times B \geq$$
$$E_1[U_{HH|M}(CF_1 = H,\ CF_2 = HH)] = \alpha\times[HH]+\beta\times[M]+ 0\times B-\pi$$

For the above to bind then,

$$\pi \geq -\alpha\times(M-HH).$$

When $CF_1 = H$ and $CF_2 \in \{HH, H, M\}$ the following must hold to ensure neither manager mimics the same type:

$$\pi\geq -\alpha\times(M-HH). \tag{A-1}$$

WHEN $CF_1 = M$

- good manager with $CF_2 = L$ does not mimic good manager with $CF_2 = H$

$$E_1[U_{L|L}(CF_1 = M,\ CF_2 = L)] = \alpha\times[L]+\beta\times[L]+\lambda_2(L)\times B \geq$$
$$E_1[U_{H|L}(CF_1 = M,\ CF_2 = H)] = \alpha\times[H]+\beta\times[L]+0\times B-\pi$$

In this case, since good and bad are observationally the same, I only need to check that a good manager will not mimic another good manager with higher cash flows.

For the above to bind then,

$$\pi\geq -\alpha\times(L-H)-\lambda_2(L)\times B. \tag{A-2}$$

WHEN $CF_1 = L$

- good manager with $CF_2 = LL$ does not mimic good manager with $CF_2 = MM$

$$E_1[U_{LL|LL}(CF_1 = L,\ CF_2 = LL)] = \alpha\times[LL+(L-M)\times(1+\underline{R}_g)] + \beta\times[LL+(L-M)\ (1+\underline{R}_g)]+1\times B \geq$$

$$E_1[U_{M|LL}(CF_1 = L,\ CF_2 = M)] = \alpha\times[M+(LL-M)\times(1-\underline{R}_g)] +\beta\times[LL + (L-M)\times(1+\underline{R}_g)]+0\times B-\pi$$

For the above to bind then,

$$\pi\geq -\alpha\times(LL-M)-1\times B.$$

- bad manager with $CF_2 = LL$ does not mimic bad manager with $CF_2 = M$

$$E_1[U_{LL|LL}(CF_1 = L,\ CF_2 = LL)] = \alpha \times [LL] + \beta \times [LL\,] + 0 \times B \geq$$

$$E_1[U_{M|LL}(CF_1 = L,\ CF_2 = M)] = \alpha \times [M] + \beta \times [LL] + 0 \times B - \pi$$

For the above to bind then,

$$\pi \geq -\alpha \times (LL - M).$$

To ensure that neither manager mimics the same type manager with a higher cash flow, the following must hold

$$\pi \geq -\alpha \times (LL - M). \quad \text{(A-3)}$$

Thus, for IC to hold for $t = 1$, one of the following must hold:

(A-1) $\pi \geq -\alpha \times (M - HH)$
(A-2) $\pi \geq -\alpha \times (L - H) - \lambda_2(L) \times B$
(A-3) $\pi \geq -\alpha \times (LL - M)$

Since H >> M, equation (A-1) is greater than (A-3). Also, since $HH - H > M - L$, equation (A-1) will be greater than equation (A-2).

Next, I look at out-of-equilibrium (o.o.e.) moves. Using the Cho-Kreps intuitive criteria, I can prune the good managers from the set of managers deviating as they are worse off from the deviation. Over the remaining set, if a manager pays a d_2 not equal to one of the equilibrium dividend levels, the manager is believed to be bad w.p. 1. He loses any reputational bonus, B, and also incurs a personal cost of π for the o.o.e. dividend level. As shown above, no one will deviate. ■

Proof of Lemma 1: Since all cash flows and costs, both distress financing and carrying, are known at the end of period 1, when the bad manager is deciding on the $t = 2$ dividend to announce at $t = 1$, d_2, he cannot mimic a good manager. He can, however, mimic a manager of the same type but with higher cash flows. That is, a bad manager with $CF_1 = H$ and $CF_2 = H$ cannot mimic a good manager with $CF_1 = H$ and $CF_2 \in \{M,H,HH\}$ when announcing d_2 but can mimic a bad manager with $CF_1 = H$ and $CF_2 = HH$. Note that bad managers with $CF_1 = M$ will not mimic good managers as they are observationally equivalent at $t = 1$, and ensuring that the bad manager does not mimic a bad manager with higher CF_2 also ensures that a bad manager will not mimic a good manager with higher CF_2.

If I look at the second period, I see that the incentive compatibility constraint is:

E_1[U(no mimic)] + E_1[U(mimic higher-CF manager of same type)].

If I examine the two period incentive compatibility constraint:

$$E_0[U(\text{no mimic})]+E_1[U(\text{no mimic})] \geq E_0[U(\text{mimic with } d \neq d1)]+E_1[U(\text{no mimic (since no mimicking is possible as all information revealed at end of period 1))}] \quad \text{(A-4)}$$

where d is the announced dividend not equal to the equilibrium dividend level d_1.

Equation (A-4) simplifies to

$$E_0[U(\text{no mimic})] \geq E_0[U(\text{mimic with higher } d_1)]. \quad \text{(A-5)}$$

This is the same as if I examined the first period separate from the second period. ■

Proof of Proposition 2: I show it is incentive compatible for the good managers and bad managers to follow the equilibrium strategies stated in Proposition 2. Note that the utility indicates the signaled dividend given the managers true type.

WHEN $CF_1 = H$

- to ensure that the bad manager will not mimic the good manager the following must hold:

$$\begin{aligned} E_0[U(d_1 = H|b)] &= \alpha\times[H+\{\delta_b^{HH}\times HH+\delta_b^{H}\times H+\delta_b^{M}\times M\}] \\ &\quad +\beta\times[H+\{\delta_b^{HH}\times HH+\delta_b^{H}\times H+\delta_b^{M}\times M\}]+0\times B \\ &\geq \quad \text{(A-6)} \\ E_0[U(d_1 = d_1^*|g)] &= \alpha\times[(H-d_1^*)\times(1-cs_g)+d_1^*+\{\delta_g^{HH}\times HH+\delta_g^{H}\times H+d_g^{M}\times M\}] \\ &\quad +\beta\times[(H-d_1^*)\times(1-cs_b)+d_1^*+\{\delta_b^{HH}\times HH+\delta_b^{H}\times H+d_b^{M}\times M\}]+0\times B \end{aligned}$$

where d_1^* is given by equation (8). Since d_1^* is solved for such that the bad manager does not mimic the good, equation (A-6) holds tightly.

- to ensure that the good manager will not mimic the bad manager the following must be true:

$$E_0[U(d_1 = d_1^*|g)] = \alpha\times[(H-d_1^*)\times(1-cs_g)+d_1^*+\{\delta_g^{HH}\times HH+\delta_g^{H}\times H+d_g^{M}\times M\}]$$
$$+\beta\times[(H-d_1^*)\times(1-cs_g)+d_1^*+\{\delta_g^{HH}\times HH+\delta_g^{H}\times H+d_g^{M}\times M\}]+1\times B$$
$$\geq \tag{A-6}$$
$$E_0[U(d_1 = H|g)] = \alpha\times[H+\{\delta_b^{HH}\times HH+\delta_b^{H}\times H+\delta_b^{M}\times M\}]$$
$$+\beta\times[H+\{\delta_g^{HH}\times HH+\delta_g^{H}\times H+\delta_g^{M}\times M\}]+0\times B$$

Equation (A-7) simplifies to proving:

$$\alpha\times[(H-d_1^*)\times(cs_b-cs_g)+(\delta_g^{HH}-\delta_b^{HH})\times HH+(\delta_g^{H}-\delta_b^{H})\times H+(\delta_g^{M}-\delta_b^{M})\times M]$$
$$+1\times B\geq 0$$

which is true since $\delta_b^{HH} < \delta_g^{HH}$, $\delta_b^{H} = \delta_g^{H}$, and $\delta_b^{M} > \delta_g^{M}$ and HH > H >> M.

WHEN CF_1 = M

Both good and bad managers follow the same strategy at this point so there is nothing for either to mimic.

WHEN CF_1 = L

- to ensure that the bad manager will not mimic the good manager the following must be true:

$$E_0[U(d_1 = L|b)] = \alpha\times[L+\{\delta_b^{M}\times M+\delta_b^{L}\times L+\delta_b^{LL}\times LL\}]$$
$$+\beta\times[L+\{\delta_b^{M}\times M+\delta_b^{L}\times L+\delta_b^{LL}\times LL\}]+0\times B$$
$$\geq$$
$$E_0[U(d_1 = M|b)] = \alpha\times[P_0^{m}(d_1 = M)]$$
$$+\beta\times[(L-M)\times(1+\underline{R}_b)+M+\{\delta_b^{M}\times M+\delta_b^{L}\times L+\delta_b^{LL}\times LL\}]+0\times B$$

This simplifies to

$$\underline{R}_b \geq \alpha\times[P_0^{m}(d_1 = M)-L-\{\delta_b^{M}\times M+\delta_b^{L}\times L+\delta_b^{LL}\times LL\}]/\beta\times(M-L). \tag{A-8}$$

- to ensure that the good manager will not mimic the bad manager the following must be true:

$$E_0[U(d_1 = M|g)] = \alpha\times[P_0^{m}(d_0 = M)]+$$
$$\beta\times[(L-M)\times(1+\underline{R}_g)+M+\{\delta_g^{M}\times M+\delta_g^{L}\times L+\delta_g^{LL}\times LL\}]+1\times B$$

$$\geq$$

$$E_0[U(d_1 = L|g)] = \alpha\times[L+\{\delta_b^{M}\times M+\delta_b^{L}\times L+\delta_b^{LL}\times LL\}]$$

$$+\beta\times[L+\{\delta_g^{M}\times M+\delta_g^{L}\times L+\delta_g^{LL}\times LL\}]+0\times B$$

This simplifies to

$$\underline{R}_g \geq (\alpha\times[P_0^{m}(d_1 = M)-L-\{\delta_b^{M}\times M+\delta_b^{L}\times L+\delta_b^{LL}\times LL\}]+B)/\beta\times(M-L). \quad \text{(A-9)}$$

With a little algebra, one can show that if (A-8) binds then (A-9) will bind since $\underline{R}_g < \underline{R}_b$ and $B > 0$.

Next, I need to prove that managers with $d_1 = M$ do not mimic good or bad managers with $CF_1 = H$. This is essentially checking to make sure that good manager with $CF_1 = M$ does not mimic a good or bad manager with $CF_1 = H$, and bad managers with $CF_1 = M$ do not mimic good or bad managers with $CF_1 = H$. (I do not verify that managers with $CF_1 = L$ since if managers with $CF_1 = M$ do not mimic, managers with $CF_1 = L$ will not.) Note that since $H \gg M$, then borrowing $M-d_1^*$ will entail a higher borrowing cost, $\overline{R}_i$. Further, the subscripts stand for the signaled and true CF_1 respectively.

- good managers with $CF_1 = M$ mimic good managers with $CF_1 = H$

$$E_0[U_{M|M}(d_1 = M)] = \alpha\times[P_0^{m}(d_1 = M)]$$

$$+\beta\times[M+\{\delta_g^{M}\times M+\delta_g^{L}\times L+\delta_g^{LL}\times LL\}]+\lambda_1\times B$$

$$\geq$$

$$E_0[U_{H|M}(d_1 = d_1^*)] = \alpha\times[(H-d_1^*)\times(1-cs_g)+d_1^*+\{\delta_g^{HH}\times HH+\delta_g^{H}\times H+\delta_g^{M}\times M\}]$$

$$+\beta\times[(M-d_1^*)\times(1+\overline{R}_g)+d_1^*+\{\delta_g^{M}\times M+\delta_g^{L}\times L+\delta_g^{LL}\times LL\}]+1\times B$$

Given the good manager with $CF_1 = M$ will need to borrow money to pay d_1^*, then the market sees the borrowing costs of $\overline{R}_g$. Therefore, the bonus will be paid in full. Further the assumption is that $d_1^* > M$. Again, the benefit of mimicking, the increase in signaled valued and the increase in the bonus, must be less than the borrowing costs;

$$\overline{R}_g \geq \frac{\alpha\times[H+\{\delta_g^{HH}\times HH+\delta_g^{H}\times H+\delta_g^{M}\times M\}-P_0^{m}(d_1 = M)]+B\times(1-\lambda_1)}{\beta\times(d_1^*-M)}. \quad \text{(A-10)}$$

- good managers with CF1 = M mimic bad managers with $CF_1 = H$

$$E_0[U_{M|M}(d_1 = M)] = \alpha\times[P_0^m(d = M)]$$

$$+\beta\times[M+\{\delta_g^M\times M+\delta_g^L\times L+\delta_g^{LL}\times LL\}]+\lambda_1\times B$$

$$\geq$$

$$E_0[U_{H|M}(d_1 = H)] = \alpha\times[H+\{\delta_b^{HH}\times HH+\delta_b^H\times H+\delta_b^M\times M\}]$$

$$+\beta\times[(M-H)\times(1+\overline{R_g})+H+\{\delta_g^M\times M+\delta_g^L\times L+\delta_g^{LL}\times LL\}]+1\times B$$

Given the good manager with $CF_1 = M$ will need to borrow money to pay $d_1 = H$ and the market will observe a borrowing costs of $\overline{R_g}$, the bonus is paid in full. This simplifies to

$$\alpha\times[P_0^m(d_0 = M)-H-\{\delta_b^{HH}\times HH+\delta_b^H\times H+\delta_b^M\times M\}]+B\times(\lambda_1-1)$$

$$-\beta\times[(M-H)\times]\overline{R_g}]\geq 0 \qquad \text{(A-11)}$$

Note that if the good manager with $CF_1 = M$ does not mimic a good manager with $CF_1 = H$, he will not mimic the bad manager, since the increase in the signaled value is smaller by mimicking the bad manager, and the cost of borrowing more is greater. That is, (A-10) > (A-11).

- bad managers with $CF_1 = M$ mimic good managers with $CF_1 = H$

$$E_0[U_{M|M}(d_1 = M)] = \alpha\times[P_0^m(d_1 = M)]$$

$$+\beta\times[M+\{\delta_b^M\times M+\delta_b^L\times L+\delta_b^{LL}\times LL\}]+\lambda_1\times B$$

$$\geq$$

$$E_0[U_{H|M}(d_1 = d_1^*)] = \alpha\times[(H-d_1^*)\times(1-cs_g)+d_1^*+\{\delta_g^{HH}\times HH+\delta_g^H\times H+\delta_g^M\times M\}]$$

$$+\beta\times[(M-d_1^*)\times(1+\overline{R_b})+d_1^*+\{\delta_b^M\times M+\delta_b^L\times L+\delta_b^{LL}\times LL\}]+0\times B$$

Given the bad manager with $CF_1 = M$ will need to borrow money to pay d_1^* and the market will observe the borrowing costs of $\overline{R_b}$, no bonus is paid. This simplifies to

$$\overline{R_b} \geq \frac{\alpha\times[H+\{\delta_g^{HH}\times HH+\delta_g^H\times H+\delta_g^M\times M\}-P_0^m(d_1 = M)]-B\times\lambda_1.}{\beta\times(H-M)} \qquad \text{(A-12)}$$

- bad managers with $CF_1 = M$ mimic bad managers with $CF_1 = H$

$$E_0[U_{M|M}(d_1 = M)] = \alpha\times[P_0^m(d_1 = M)]$$

$$+\beta\times[M+\{\delta_b^M\times M+\delta_b^L\times L+\delta_b^{LL}\times LL\}]+\lambda_1\times B$$

$$\geq$$

$$E_0[U_{H|M}(d_1 = H)] = \alpha\times[H+\{\delta_b^{HH}\times HH+\delta_b^H\times H+\delta_b^M\times M\}]$$

$$+\beta\times[(M-H)\times(1+\overline{R_b})+H+\{\delta_b^M\times M+\delta_b^L\times L+\delta_b^{LL}\times LL\}]+0\times B$$

Given the bad manager with $CF_1 = M$ will need to borrow money to pay $d_1 = H$ and the market will observe the borrowing costs of $\overline{R_b}$, no bonus is paid. This simplifies to

$$\alpha\times[P_0^m(d_1 = M)-H+\{\delta_b^{HH}\times HH+\delta_b^H\times H+\delta_b^M\times M\}]$$

$$+\lambda_1\times B-\beta\times[(M-H)\overline{R_b}]\geq 0 \qquad \text{(A-13)}$$

However, if a bad manager with $CF_1 = M$ does not mimic a good manager with $CF_1 = H$, he will not mimic a bad manager with $CF_1 = H$ since the benefit from mimicking a bad manager with $CF_1 = H$ is less than mimicking a good manager with $CF_1 = H$ and the costs are greater. That is, (A-12) > (A-13).

Further, one can show that (A-10) > (A-12). Thus, the binding constraints that must hold for the parameter values of $\underline{R}_b$ and $\overline{R}_g$ given by (A-8) and (A-10), respectively.

Next, I look at o.o.e. moves. Using Cho-Kreps Intuitive Criteria, I can prune the good managers with high and low cash flows from deviating as their utility will be lower. Over the remaining set, if a manager pays a d_1 not equal to one of the equilibrium dividend levels, the manager is believed to be bad w.p. 1. He will lose the reputational bonus, B, and as shown above, no one will deviate. ■

Proof of Proposition 3: Good managers with $CF_1 = H$ pay $d_1^* < H$, and good managers with $CF_1 = L$ pay $d_1 = M > L$. Thus, good managers are smoothing dividends relative to bad managers. Further, both of these dividend levels, d_1^* and $d_1 = M$, are closer to $d_0 = M$ than H or L. Also, for good managers the variation in dividends is less than the variation in earnings. ■

NOTES

1. See Lintner (1956).
2. See Lintner (1956) and Fama and Babiak (1968)
3. See, for example, Bhattacharya (1979), John and Williams (1985), Miller and Rock (1985), and Ofer and Thakor (1987).
4. Though this is a strong assumption, most exchange listed firms announce their quarterly earnings and also have analysts following their performance. Further, as Marsh and Merton (1987) show, one can determine the permanent earnings component from reported earnings. Both of these facts lead to the market verifying the firm's earnings. In contrast, managerial type is unknown and often impossible to verify.
5. For computation ease, I assume that $HH-H > M-L$.
6. Since I want to focus on intertemporal dividend smoothing, I assume an existing dividend level.
7. The results of the model would still hold if the good manager's carrying costs were positive; $CSg > 0$.
8. To motivate this assumption, I assume that banks have some special ability to determine firm cash flows. For example, Leland and Pyle (1977), Diamond (1984), and Ramakrishnan and Thakor (1984), show that financial intermediaries resolve information problems. James (1987) empirically shows that borrowers experience abnormal returns when they announce bank loans. Thus, banks have an advantage at determining creditor quality. Therefore, in the context of this paper, the bank is able to determine firm cash flows, offer a better rate to the good manager, and once the bank loan is made, firm cash flows are known.
9. Benartzi, Michaely, and Thaler (1997) suggest that dividends do not signal future CFs but rather say something about current earnings. However, dividends are often decided upon and announced prior to the earnings realization. Thus, these announcements are signaling future earnings but are paid concurrent to when earnings are realized by the firm and known by the market. Further, Garrett and Priestley (2000) find a strong positive relation between current dividend changes and both current and future permanent earnings.
10. The linear utility function of the manager is assumed similar to Holmstrom and Milgrom (1991) and Miller and Rock (1985).
11. For simplicity, I assume that π is a fixed cost.
12. $\lambda_2(L) = \theta\times\delta_g^{\,L}/[\theta\times\delta_g^{\,L}+(1-\theta)\times\delta_b^{\,L}] < \lambda_1 = \theta$, $\lambda_2(H) = \theta\times\delta_g^{\,H}/[\theta\times\delta_g^{\,H}+(1-\theta)\times\delta_b^{\,H}] > \lambda_1 = \theta$, and $\lambda_2(M) = \lambda_1 = \theta$.
13. For example, $\mu_g^{\,M} = \delta_g^{\,M}/(\delta_g^{\,M}+\delta_b^{\,M}+\delta_g^{\,L})$, and $E_g[CF_2^{\,M}(d_1)] = \delta_g^{\,H}\times H+\delta_g^{\,M}\times M+\delta_g^{\,L}\times L$.
14. I assume that d_1^*-M is such that a manager who promises d_1^* but has CF_1 less than H will need outside financing at the higher marginal rate. Otherwise a good manager with $CF_1 = M$ will mimic a good manager with $CF_1 = H$.
15. See, for example, Aharony and Swary (1980), Asquith and Mullins (1983), Bhattacharya (1979), Eades, Hess, and Kim (1985), John and Williams (1985), Kalay and Loewenstein (1985, 1986), Miller and Rock (1985), and Ofer and Thakor (1987).
16. Lee, Mucklow, and Ready (1993) find that information asymmetry around an earnings announcement for an NYSE firm is quickly resolved in a matter of a few hours. Thus, I choose a three-day window with the desire to eliminate biases arising from other information events.

17. I would like to thank Todd Milbourn for supplying the CEO tenure data.
18. *Value Line* predictions from 1982 through the announcement year were used to determine if there existed any bias in these predictions. Each year's quarterly dividends from CRSP were summed and the difference between the actual (CRSP) and forecasted (*Value Line*) dividend was calculated. This difference was summed over seven years to find any error. None of the firms in the sample had significant forecast error.
19. The standard deviation in the ratio of quarterly dividends to quarterly earnings, the variance of dividends, and the number of times a firm has an unexpected dividend change over the previous three years were all used as measures of dividend smoothing. Results were analogous to those for the ratio of the standard deviation in quarterly dividends to the standard deviation of quarterly earnings, thus, these results are omitted for the sake of brevity.
20. In the weighted least squares regressions the reciprocal of the firm's forecast variance is used as the weight.
21. See Fama and French (2000) for a discussion of the characteristics of firms that pay dividends.
22. To control for the fact that some firms with large variations in earnings and dividends will have a similar smoothing measure as firms with little variation in earnings and dividends, I divided the sample into quartiles based on past earnings, and run the regression in equation (6) for each quartile. Results remain qualitatively the same for each quartile as for the entire sample.
23. The regression in equation (6) was also performed separately on unexpected dividend increases and unexpected dividend decreases. The results, aside from the coefficient on firm size, are qualitatively the same as what is reported.
24. I use the par value of preferred stock if the market value is unavailable.
25. See Allen and Michaely (1995).

ACKNOWLEDGMENTS

I would like to thank Jeff Bacidore, Alex Butler, Rob Bliss, Shane Corwin, Jeff Green, Craig Holden, Robert Hauswald, Sreenivas Kamma, Marc Lipson, Richard Shockely, Scott Smart, Anjan Thakor, and seminar participants at the University of Georgia, the 1998 EFA meetings, and the 2000 FMA meeting, for their comments. Any remaining errors or omissions are mine.

REFERENCES

Aharony, J., & Swary, I. (1980). Quarterly dividend and earnings announcements and stockholders' returns: An empirical analysis. *Journal of Finance*, *35*, 1–12.

Allen, F., & Michaely, R. (1995). Dividend policy. In: R. A. Jarrow, V. Maksimovic & W. T. Ziemba (Eds), *Handbooks in Operations Research and Management Science: Finance* (Vol. 9). Amsterdam, Elsevier.

Asquith, P., & Mullins, Jr., D. (1983). The impact of initiating dividend payments on shareholders' wealth. *Journal of Business*, *56*, 77–96.

Bajaj, M., & Vijh, A. (1990). Dividend clienteles and the information content of dividend changes. *Journal of Financial Economics, 26*, 193–220.

Benartzi, S., Michaely, R., & Thaler, R. (1997). Do changes in dividends signal the future or the past? *Journal of Finance, 52*, 1007–1034.

Bhattacharya, S. (1979). Imperfect information, dividend policy, and "the bird in the hand" fallacy. *Bell Journal of Economics, 10*, 259–270.

Brown, S., & Warner, J. (1980). Measuring security price performance. *Journal of Financial Economics, 8*, 205–258.

Brown, S., & Warner, J. (1985). Using daily stock returns: The case of event studies. *Journal of Financial Economics, 14*, 3–31.

Chowdhry, B., & Nanda, V. (1994). Repurchase premia as a reason for dividends: A dynamic model of corporate payout policies. *Review of Financial Studies, 7*, 321–350.

Denis, D., Denis, D., & Sarin, A. (1994). The information content of dividend changes: Cash flow signaling, overinvestment, and dividend clienteles. *Journal of Financial and Quantitative Analysis, 29*, 567–587.

Diamond, D. (1984). Financial intermediation and delegated monitoring. *Review of Economic Studies, 51*, 393–414.

Eades, K., Hess, P., & Kim, E. H. (1985). Market rationality and dividend announcements. *Journal of Financial Economics, 14*, 581–604.

Eberhart, A., & Damodaran, A. (1997). *Relative valuation, differential information, and cross-sectional differences in stock return volatility*. Georgetown University, Working Paper.

Fama, E., & Babiak, H. (1968). Dividend policy: An empirical analysis. *Journal of American Statistical Association, 63*, 1132–1161.

Fama, E., & French, K. (2000). *Disappearing dividends: Changing firm characteristics or lower propensity to pay?* University of Chicago, Working Paper No. 509.

Fudenberg, D., & Tirole, J. (1995). A theory of income and dividend smoothing based on incumbency rents. *Journal of Political Economy, 103*, 75–93.

Garrett, I., & Priestley, R. (2000). Dividend behavior and dividend signaling. *Journal of Financial and Quantitative Analysis, 35*, 173–189.

Holmstrom, B., & Milgrom, P. (1991). Multitask principal-agent analyses: Incentive contracts, asset ownership, and job design. *Journal of Law, Economics and Organization, 7*, 24–52.

James, C. (1987). Some evidence on the uniqueness of bank loans. *Journal of Financial Economics, 19*, 217–235.

Jensen, M. (1986). Agency costs of free cash flow, corporate finance, and takeovers. *American Economic Review, 76*, 323–329.

John, K., & Williams, J. (1985). Dividends, dilution, and taxes: A signaling equilibrium. *Journal of Finance, 40*, 1053–1070.

Juster, A. (1996). Signaling behavior and dividend smoothing: A dynamic model. Tulane University, Working Paper.

Kalay, A., & Loewenstein, U. (1985). Predictable events and excess returns: The case of dividend announcements. *Journal of Financial Economics, 14*, 423–449.

Kalay, A., & Loewenstein, U. (1986). The informational content of the timing of dividend announcements. *Journal of Financial Economics, 16*, 373–388.

Kumar, P. (1988). Shareholder-manager conflict and the information content of dividends. *Review of Financial Studies, 1*, 111–136.

Laub, P. (1976). On the informational content of dividends. *Journal of Business, 49*, 73–80.

Lee, C., Mucklow, B, & Ready, M. (1993). Spreads, depths, and the impact of earnings information: An intraday analysis. *Review of Financial Studies, 6*, 345–374.

Leland, H. & Pyle, D. (1977). Informational asymmetries, financial structure, and financial intermediation. *Journal of Finance*, *32*, 371–387.

Lintner, J. (1956). Distribution of incomes of corporations among dividends, retained earnings and taxes. *American Economic Review*, *46*, 97–113.

Marsh, T., & Merton, R. (1987). Dividend behavior for the aggregate stock market. *Journal of Business*, *60*, 1–40.

Miller, M., & Rock, K. (1985). Dividend policy under asymmetric information. *Journal of Finance*, *40*, 1031–1051.

Ofer, A., & Thakor, A. (1987). A theory of stock price responses to alternative corporate cash disbursement methods: Stock repurchases and dividends. *Journal of Finance*, *42*, 365–394.

Ramakrishnan, R., & Thakor, A. (1984). Information reliability and a theory of financial intermediation. *Review of Economic Studies*, *51*, 415–432.

Stulz, R. (1990). Managerial discretion and optimal financing policies. *Journal of Financial Economics*, *26*, 3–27.

Warther, V. (1996). *Dividend smoothing: A sleeping dogs explanation*. University of Michigan, Working Paper.

THE ROLE OF FEDERAL LAW ENFORCEMENT ACTIONS IN CORPORATE GOVERNANCE

Mark Hirschey and Elaine Jones

ABSTRACT

Large negative stock-price reactions are tied to public announcements regarding informal and formal law enforcement actions taken by the Department of Justice (DOJ), Federal Trade Commission (FTC), and the Securities and Exchange Commission (SEC). Such influences appear especially large for firms with significant reputational capital and growth opportunities. These findings suggest that federal law enforcement plays an important role in monitoring management and represents a powerful corporate governance mechanism. These findings also suggest that public announcements of federal law enforcement actions constitute a flexible tool used to improve suboptimal performance in dynamic competitive environments.

1. INTRODUCTION

A large and growing corporate governance literature builds upon the work of Jensen and Meckling (1976) who contemplate the role of corporate control mechanisms as means for helping ameliorate the potential divergence of interests

Advances in Financial Economics, Volume 6, pages 117–141.
Copyright © 2001 by Elsevier Science B.V.
All rights of reproduction in any form reserved.
ISBN: 0-7623-0713-7

between managers and stockholders. Jensen and Meckling (1976) describe how a variety of monitoring mechanisms inside and outside the firm work together to establish an optimal set of restrictions on firm activity, and why firms themselves often suggest such restrictions. Commercial bank loan covenants, financial audits by independent auditors, and performance scrutiny by independent security analysts are all common examples of outside monitoring mechanisms agreed to by firms. Other outside monitoring mechanisms can be mandatory. Such compulsory mechanisms include the wide variety of federal, state and local laws and regulations that govern corporate behavior. As Fama and Jensen (1983) point out, the potential exploitation of stockholders, bondholders and other residual claimants by opportunistic decision agents is often reflected in arguments leading to the establishment of broad regulatory initiatives, such as those stemming from establishment of the Securities and Exchange Commission (SEC).

Calculated or inadvertent violations of federal laws have the potential to impose significant costs on shareholders and other residual claimants. The pursuit of illegal short-term strategies can represent a form of self-dealing by managers who seek to reap short-term personal gain while escaping detection. Actual or suspected violations of federal laws have the potential to result in significant costs measured in terms of investigation expenditures, litigation expenses, fines and seizures, and lost reputational capital for the firm – all of which can measurably reduce future cash flows and current market values. Within this context, federal laws can be seen as part of the institutional framework that contributes to the range of control mechanisms that originate inside and outside the firm to comprise an effective system of corporate governance. Because short-term "hit and run" managers may possess incentives to "cut" legal and ethical corners, the design and administration of federal laws can be seen as a means of outside monitoring designed to ensure a coincidence of managerial incentives, stockholder interests, and broader social objectives.

The aim of this study is to investigate the role of federal law enforcement activities as an important element in the institutional framework of corporate governance. This study provides the first large-sample analysis of the stock-market reaction to the law enforcement activities of three federal agencies: the Department of Justice (DOJ), Federal Trade Commission (FTC), and SEC. Evidence of stock market wealth effects is presented for a variety of enforcement activities, including: informal investigations, formal investigations, lawsuits and settlements. Evidence of this nature is of interest because it documents the extent to which announcements regarding the enforcement activity of the executive branch and federal agencies represent a source of important new information for capital markets. Such evidence is also important because it demonstrates the role of actual or threatened legal action in corporate governance and as a tool for competition policy in dynamic markets.

Economic effects of federal law enforcement actions on the value of the firm can be expected to differ across firms according to the relative importance of intangible factors in firm valuation. Using the Myers (1977) framework, one could think of firm value as derived from assets in place and growth opportunities stemming from research and development (R&D), advertising, and other such expenditures. This study investigates the possibility that firms may be especially susceptible to federal law enforcement actions when adverse reputational and growth-option consequences are large.

The study is organized as follows. Section 2 offers background on the administrative responsibilities of the DOJ, FTC and the SEC, and summarizes prior research in the area. Section 3 describes the sample and empirical methodology employed. Empirical results are discussed in Section 4. A summary and implications for further research are provided in Section 5.

2. FEDERAL LAW ENFORCEMENT ACTIONS

2.1 Administration of Federal Laws and Regulations

Federal and state statutes govern a broad range of corporate activities, including, for example, the registration, offering, and sale of securities. The design and administration of these statutes play a key role in determining the institutional framework and competitive environment of securities markets. In terms of administration, formidable roles are played by three federal bodies: the DOJ, FTC, and SEC.

Chief purposes of the DOJ, an executive department headed by the Attorney General, are to enforce federal laws, furnish legal counsel in federal cases, and to construe the laws under which other departments act. The FTC is an independent agency created by Congress in 1914 and given broad powers to promote free and fair competition in interstate commerce through the prevention of trade restraints such as price-fixing, false advertising, boycotts, illegal combinations of competitors, and other unfair methods of competition. The FTC is responsible for much of the substantive regulation in the legal environment of business, including antitrust where duties are shared with the DOJ.

Major federal acts administered by the SEC include the Securities Act of 1933, the Securities Exchange Act of 1934, the Public Utility Holding Act of 1935, the Trust Indenture Act of 1939, the Investment Adviser's Act of 1940, and the Investment Company Act of 1940. The Securities Act of 1933 provides for the registration of securities which are to be sold to the public and for complete information as to the issuer and the stock offering. The Securities Exchange Act of 1934 created the SEC, an independent administrative agency with five commissioners, and governs the operation of stock exchanges and over the counter trading, and the publication of information concerning stocks listed on these exchanges. The SEC has broad rule-making authority. Formal rules carry the force of law; informal communications clarify and interpret current enforcement strategies and viewpoints (see Butler, 1987).

In practice, DOJ, FTC and SEC staff investigations are often informal at first, and terminated if insufficient evidence exists to charge the target of the investigation with a violation of federal laws. Alternatively, staff investigations may be terminated if investigation targets agree to stop engaging in the allegedly unfair or deceptive practice. Many, if not most, informal investigations can be settled with limited fanfare. For example, if the FTC determines that there is adequate basis for bringing a formal complaint, defendants can settle their dispute by agreeing to the terms of a consent decree. Consent decrees often contain the terms of the settlement, including redress for injured consumers, payment of penalties, and prohibition of certain practices. If the defendant and the FTC cannot agree on a settlement, then the matter goes to trial before an administrative law judge who may dismiss the complaint or issue a cease and desist order. In some instances, the FTC may file a lawsuit in federal district court, perhaps with the support of the DOJ and/or SEC, to seek an injunction against some disapproved of practice while it is being challenged in FTC proceedings (see Butler, 1987).

2.2 Valuation Effects of Enforcement Actions

In a fully informed stock market, news regarding the enforcement of federal laws against publicly traded firms would have no effect on target firm stock prices. A "rational expectation's hypothesis" predicts that investors would be unaffected by announcements concerning the enforcement actions of federal agencies because current stock prices accurately reflect discounted future cash flows based upon all relevant information. An absence of announcement effects tied to federal law enforcement actions would suggest that the market is fully aware of illegal activity, the probability of getting caught, and the potential sanctions tied to detection and conviction. Future cash flows lost following federal law enforcement actions can include the costs of sacrificing illegal advantages over competitors, investigation expenditures, litigation expenses, and lost reputational capital. An absence of abnormal returns tied to federal law enforcement actions does not mean that there is no cost to being caught; it simply implies that the market correctly anticipates the magnitude and probability of such costs. Here it is important to recognize that the term "caught" does not necessarily imply guilt as well. Under the rational expectation's hypothesis, market participants also know the probability of innocent firms being investigated or sued.

Positive abnormal returns associated with public announcements regarding the enforcement actions of federal agencies could lead to different conclusions regarding market efficiency and enforcement efficacy. Positive abnormal returns could suggest that news regarding the enforcement process represents favorable new information for investors. Positive abnormal returns could simply represent "good news" for target firms in that enforcement sanctions are more narrow than generally expected. Alternatively, prior to public announcements surrounding enforcement activity, government agencies might possess information that is relevant to the estimation of future cash flows for target firms. While such good information is not reflected in the pre-enforcement activity announcement price, prices adjust favorably in the post-announcement period.

Negative abnormal returns tied to announcements concerning federal law enforcement actions would also suggest that news regarding the enforcement process represents a source of new information for investors. That is, prior to *The Wall Street Journal* announcements, federal agencies possess damaging information that is relevant to a target firm's future cash flows, but is not reflected in its pre-enforcement announcement price. Why such news is "bad" is unclear. Announcements in *The Wall Street Journal* regarding enforcement activities might reveal information about illegal behavior not previously known by the market. On the other hand, "guilty" firms whom the market did not believe would be caught may now be the recognized target of costly enforcement sanctions. Alternatively, "innocent" firms who have not engaged in any illegal behavior are now subject to enforcement activities. In both instances, the direct and indirect costs tied to federal law enforcement actions could be expected to lead to a negative stock-market response during the announcement period.

2.3 Prior Studies

A few prior studies have been conducted to test limited aspects of the stock-market reaction to enforcement activity of the executive branch and federal agencies. However, none of these studies are as comprehensive as the one presented here, nor do they address the role of federal regulatory agencies in monitoring managerial behavior.

For example, in an early study, Garbade, Silber and White (1982) investigate the possibility of abnormal returns for a sample of 34 firms resulting from antitrust lawsuits filed by the FTC or DOJ. Garbade, Silber and White (1982) also conduct a cross-sectional analysis comparing the magnitude of abnormal returns with the fraction of the firm's sales revenue originating under the area of attack, the financial position of the firm, and a dummy variable for pre- or post-dating of a landmark case in the acceptance of treble damages for antitrust violations. Garbade, Silber and White (1982) find a statistically significant drop of −5.87% (t-statistic = −7.71) in the value of the stock of these companies within four trading days (0,+3) after the filing of a DOJ or FTC suit. After day (4), Garbade, Silber and White (1982) find no mean abnormal returns less than zero. In terms of cross-sectional results, Garbade, Silber and White (1982) find that the magnitude of the share-price response tied to DOJ and FTC lawsuit announcements is related to variables that indicate the likely effects of such suits on future earnings.

Feroz, Park and Pastena (1991) consider the stock-market response to the disclosure of accounting violations and subsequent SEC enforcement actions. These authors focus on stock-price effects associated with news of the alleged reporting violation, associated SEC investigation, and resulting settlement. They also offer a cross-sectional analysis of the relationship between the magnitude of abnormal returns and the income effect of the disputed accounting convention, the type of accounting ledger item in question, and the possibility that such a reporting error is fraud-related. Feroz, Park and Pastena (1991) found that reports of disclosure violations had a statistically significant average two-day (−1, 0) abnormal return of −12.9% (t-statistic = −3.75) for 58 firms. Even for a subsample of 20 firms that had previously disclosed the effect of some disputed accounting on earnings, the subsequent announcement of a SEC investigation led to a statistically significant two-day abnormal return of −6.0% (t-statistic = −2.38). This means that the market reacts negatively to announcements of a SEC investigation, even when there is prior specific knowledge of the accounting errors involved. Such incremental effects of SEC investigation announcements on shareholder wealth may stem from the negative publicity generated and the possibility of third-party lawsuits. At a minimum, Feroz, Park and Pastena (1991) argue that the ability of SEC investigation announcements to affect the share prices of target firms suggests that the agency possesses a potent sanction that gives managers a market-based incentive to avoid investigation. Interestingly, Feroz, Park and Pastena (1991) report no significant abnormal returns tied to the announcement of settlements of SEC investigations. Apparently, the market discounts such favorable news well in advance of the formal announcement.

And finally, other recent studies of interest include Bittlingmayer (1993) and Bhagat, Brickley and Coles (1994). Bittlingmayer argues that increased government antitrust enforcement is related to declines in the broad market indexes.

Moreover, Bittlingmayer provides evidence of abnormally negative monthly returns for the common stock of 11 firms sued by the federal government between January 1904 and July 1914. For these 11 firms, negative abnormal returns of -3.47% (t-statistic = -11.47) exist over the two-year period surrounding the month in which such lawsuits are filed. Bittlingmayer argues that the existence of negative abnormal returns prior to filing dates can be explained by the fact that government antitrust filings are anticipated over a period of several months, on average, and are typically preceded by costly investigations. Bhagat, Brickley and Coles (1994) conduct an analysis of the stock-market reaction to interfirm litigation. Event study results show that the combined wealth effect for matched pairs of plaintiffs and defendants is a negative and statistically significant -1.04% (z-statistic = -3.56). Defendants see a significant drop in their stock price of -0.92% (z-statistic = -4.18); the effects on plaintiff stock prices are a more muted and statistically insignificant -0.18% (z-statistic = -0.48). Importantly, plaintiffs do not gain by the same amount as defendants lose. Bhagat, Brickley and Coles (1994) conclude that bargaining among firm claimants sometimes leads to inefficient outcomes because of the costs of increased financial distress imposed on defendants. The Bhagat, Brickley and Coles (1994) study is of interest here because it represents the most comprehensive study of the wealth-effects of private litigation. Their study creates a structure for thinking about how the cost and benefits of private litigation are distributed. The intent of this study is to create a similar foundation for the analysis of litigation tied to the enforcement actions of public agencies.

3. DATA AND METHODOLOGY

3.1 The Sample

This study of enforcement actions against firms is restricted to those taken by the DOJ, FTC and SEC during the 1992–1995 four-year period. Event dates for enforcement action announcements are identified from *The Wall Street Journal Index* on-line (*WSJI*). The *WSJI* is an attractive source for event date information because it offers a precise indication of when the stock market first received relevant news regarding specific law enforcement actions. It is extremely difficult to identify when the market receives details regarding investigations, suit filings, and settlements when they are reported in original sources such as SEC filings or official court documents. Following Feroz, Park and Pastena (1991), we typically identify event-day (0) as the announcement date in *The Wall Street Journal*. In 23 of 163 instances, *The Wall Street Journal* announcement specifically indicated that a company press release had been made on the prior day. In these 23 cases, the company's press release disclosure date was identified as event-day (0).

Searches were conducted using the key words "department of justice" *or* "SEC" *or* "FTC" *and* "investigation" *or* "suit" *or* "criminal". To be included in the sample, the common stock of each firm against which enforcement action was taken must be listed on either the New York Stock Exchange(NYSE), the American Stock Exchange (AMEX), or NASDAQ, and included on the *Center for Research for Security Prices (CRSP)* daily stock returns file for six months prior to the enforcement action announcement. Firms also need to be continuously listed over the estimation and event periods to be included. This method of sample selection permits the collection of a broad and public base of evidence concerning the stock-price reaction associated with the enforcement of federal laws by the DOJ, FTC and SEC.

A typical example of an informal investigation announcement is provided in the case of securities dealer Merrill Lynch & Co. where:

> Sources say that the SEC and the securities regulators of Florida, Virginia and New Mexico are looking into the sales practices of Merrill Lynch & Co in the 1980s with regard to partnerships . . . (*The Wall Street Journal*, January 27, 1994, C1).

Sometimes, announcements of federal law enforcement actions involve a number of companies. A formal investigation announcement involving Goodyear Tire & Rubber Co., among others, is a common illustration:

> The Justice Department said on August 23, 1995 that it is investigating potential price-fixing among American tire makers. Several tire companies, including Goodyear Tire & Rubber Co, Bridgestone/Firestone Inc., Michelin North America, Dunlop Tire Corp. and Cooper Tire & Rubber Co, admitted that they had submitted documents in response to a Justice Department subpoena . . . (*The Wall Street Journal*, August 24, 1995, A3).

A standard suit filing announcement involving two companies, here software makers Microsoft Corp. and Intuit Inc., is:

> The Justice Department filed an antitrust suit to block Microsoft Corp.'s $2.1 billion acquisition of Intuit Inc., arguing that the acquisition would decrease competition and innovation in the market for personal-finance software . . . (*The Wall Street Journal*, April 28, 1995, A3).

And finally, an ordinary settlement announcement, this in the case of the insurer Presidential Life Corp., is:

> As part of a settlement with the SEC, Presidential Life Corp., principally an issuer of annuity contracts through its Presidential Life Insurance Co. unit, will restate its financial results for several years to reflect a markdown of some junk securities in its portfolio . . . (*The Wall Street Journal*, March 2, 1993, B4).

Complete data could be obtained on a sample of $N = 13$ informal investigations, $N = 97$ formal investigations, $N = 28$ lawsuit filings, and $N = 24$ settlements. Information effects are measured and reported for all these $N = 162$ law enforcement actions.

3.2 Estimation Method

The event study methodology of James (1987) is used to obtain estimates of abnormal stock returns surrounding announcements in *The Wall Street Journal* concerning various enforcement actions taken by the DOJ, FTC and SEC. The OLS market model is employed to calculate the abnormal return or prediction error for the common stock of firm j on day t, such that:

$$PE_{jt} = R_{jt} - (\alpha_j + \beta_j R_{mt}), \tag{1}$$

where R_{jt} = rate of return on the common stock of the jth firm on day t, R_{mt} = rate of return of the CRSP value-weighted market index over period t, α_j = OLS estimate of the intercept, and β_j = OLS estimate of the slope parameter that measures the sensitivity of R_{jt} to the market index.

A 180-day estimation period is used that begins 225 trading days before the event date, $t = -225$, and ends 46 trading days before the event date, $t = -46$. The event date, $t = 0$, is typically assumed to be *The Wall Street Journal* announcement date, but may be the prior day if an earlier announcement on that day is specifically noted in *The Wall Street Journal* article. Daily prediction errors are averaged over the sample of N firms yielding the average prediction errors (average abnormal returns):

$$\frac{APE_t = \sum_{j=1}^{N} PE_{jt}}{N} \tag{2}$$

$$\frac{CAPE_{T_1T_2} = \sum_{j=1}^{N} \sum_{t=T_1}^{T_2} PE_{jt}}{D_j - 2}. \tag{3}$$

Cumulative average prediction errors (cumulative average abnormal returns) are calculated over event interval periods (T_1,T_2) of 2 days (–1, 0) and (0, +1), of 3 days (−1, +1) and (0, +2), and of 4 days (−1, +2) and (0, +3) as: Equation (3) A t-test is applied to examine the hypothesis that the $CAPE_{T_1,T_2}$ are not significantly different from zero. Under the null hypothesis, each PE_{jt} has a mean zero and constant variance equal to the portfolio variance of the APE_t. The estimated standard deviation of the APE_t is:

$$s_{APE} = \sqrt{\frac{\sum_{t=-225}^{-46} (APE_t - \overline{APE})^2}{D_j - 2}} . \tag{4}$$

$\overline{APE}$ is the mean of the average prediction errors, where this mean is defined as:

$$\frac{\overline{APE} = \sum_{t=-225}^{-46} APE_t}{D_j}, \tag{5}$$

and D_j equals the number of non-missing trading days in the estimation period. The portfolio test-statistic for the average error on any day is:

$$t = \frac{APE_t}{s_{APE}} . \tag{6}$$

Assuming time-series independence, the test-statistic for each $CAPE_{T_1, T_2}$ is:

$$t = \frac{CAPE_t}{s_{APE} \sqrt{T_2 - T_1 + 1}} . \tag{7}$$

Since the test-statistic standard deviation is calculated using the time series of portfolio returns, the potential problem of cross-sectional correlation among individual security returns is avoided.

Because there are only a small number of observations for certain event types, median CAPEs and nonparametric statistics for each announcement type are also reported. Following Corrado (1989), a *z*-statistic is developed to test whether median CAPEs are different from zero. Using the daily prediction errors for firm *j* in both the 180-day estimation period (day $t = -225$ to $t = -46$) and the five-day event period (day $t = -1$ to $t = +3$), a rank from 1 to 185 is assigned to each daily prediction error. Accordingly, the rank for firm *j's* prediction error on day *t*, K_{jt}, is given by:

$$K_{jt} = rank(PE_{jt}). \tag{8}$$

The smallest prediction error for firm j will have rank $K_{jt} = 1$. The next smallest prediction error for firm j will have rank $K_{jt} = 2$, and so on; the largest prediction error will have rank $K_{jt} = 185$.

The median rank for each firm j is given by:

$$\overline{K} = \frac{185 + 1}{2} = 93. \tag{9}$$

The average rank across all firms at time t is:

$$\overline{K}_t = \frac{1}{n} \sum_{j=1}^{n} K_{jt} . \tag{10}$$

The average rank across all firms for the event window T_1 to T_2 is:

$$\overline{K_{T_1,T_2}} = \frac{1}{T_2 - T_1 + 1} \sum_{t=T_1}^{T_2} \overline{K}_t . \tag{11}$$

The rank z-statistic for the event window T_1 to T_2 is:

$$z = (T_2 - T_1 + 1)^{1/2} \left\{ \frac{\overline{K_{T_1,T_2}} - 93}{\left[\frac{1}{185} \sum_{t=1}^{185} (\overline{K}_t - 93)^2 \right]^{1/2}} \right\} \tag{12}$$

4. EMPIRICAL RESULTS

4.1 Announcement Effects by Type of Enforcement Activity

Table 1 summarizes market model cumulative average prediction errors (CAPEs) for all firms subject to federal law enforcement announcements in *The Wall Street Journal* during the four-year 1992–1995 period. CAPEs are categorized based upon the nature of enforcement activity by the DOJ, FTC and SEC, including: informal investigations, formal investigations, lawsuits filed, and settlements. Of primary interest is the wealth effect realized over the

Table 1. Cumulative Average Prediction Errors (CAPEs) for Enforcement Acitivities of the DOJ, FTC, and SEC.

Market-model cumulative average prediction errors (CAPEs), t-statistics, and z-statistics are shown for a variety of event-period windows for each type of enforcement activity. Parametric and non-parametric statistics are shown, as are the size and statistical significance of mean (average) and median differences in announcement effects for various enforcement activities. The statistical significance of mean differences are evaluated using a t-test; median differences are evaluated using a z-test.

Event-period Window	CAPEs	t-statistic	Event-period Window	Median CAPEs	z-statistic
A. Informal Investigations (N = 13)					
(−1,0)	−5.96%	−4.96 [c]	(−1,0)	−1.77%	−0.30
(−1,+1)	−4.47%	−3.04 [c]	(−1,+1)	−0.10%	−0.30
(−1,+2)	−4.25%	−2.50 [c]	(−1,+2)	−0.77%	−1.41 [a]
(0,+1)	−0.74%	−0.62	(0,+1)	0.20%	1.37 [a]
(0,+2)	−0.52%	−0.36	(0,+2)	−0.19%	−0.30
(0,+3)	0.31%	0.19	(0,+3)	1.05%	1.37 [a]
B. Formal Investigations (N = 97)					
(−1,0)	−2.34%	−5.94 [c]	(−1,0)	−1.17%	−3.13 [c]
(−1,+1)	−2.06%	−4.28 [c]	(−1,+1)	−1.84%	−3.13 [c]
(−1,+2)	−2.24%	−4.01 [c]	(−1,+2)	−2.17%	−2.93 [c]
(0,+1)	−1.99%	−5.04 [c]	(0,+1)	−1.34%	−2.93 [c]
(0,+2)	−2.16%	−4.47 [c]	(0,+2)	−1.64%	−2.12 [b]
(0,+3)	−2.19%	−3.93 [c]	(0,+3)	−1.07%	−2.32 [b]
C. Suits Filed (N = 28)					
(−1,0)	−1.19%	−1.58 [a]	(−1,0)	−1.19%	−0.53
(−1,+1)	−1.43%	−1.55 [a]	(−1,+1)	−0.98%	−1.67 [b]
(−1,+2)	−1.01%	−0.95	(−1,+2)	−1.16%	−0.91
(0,+1)	−1.41%	−1.88 [b]	(0,+1)	−1.11%	−2.42 [c]
(0,+2)	−0.99%	−1.07	(0,+2)	−1.08%	−0.91
(0,+3)	−0.85%	−0.80	(0,+3)	−1.00%	0.22
D. Settlements (N = 24)					
(−1,0)	1.56%	1.14	(−1,0)	−0.33%	−0.29
(−1,+1)	1.63%	0.97	(−1,+1)	0.74%	1.77 [b]
(−1,+2)	1.10%	0.57	(−1,+2)	0.34%	0.54
(0,+1)	−0.23%	−0.17	(0,+1)	0.15%	0.95
(0,+2)	−0.76%	−0.45	(0,+2)	−0.53%	−0.70
(0,+3)	−1.03%	−0.53	(0,+3)	−1.47%	−1.11

[a] Significant at the 10% level (one-tail test).

[b] Significant at the 5% level (one-tail test).

[c] Significant at the 1% level (one-tail test).

(−1, 0) event period. However, to test for the robustness of (−1, 0) results, CAPEs are reported for five additional event-period windows, including: (−1, +1), (−1, +2), (0, +1), (0, +2), (0, +3).

On average, informal investigations have statistically significant negative wealth effects over the (−1,0) event window of −5.96% ($t = -4.96$). Negative wealth effects appear to be fairly robust for informal investigations in that statistically significant results are also found over the (−1,+1) and (−1,+2) event periods. Nonparametric results provide corroborating evidence in that statistically significant negative wealth effects of −0.77% (z-statistic = −1.41) are noted over the (−1,+2) event period. Paradoxically, nonparametric results suggest weak but statistically significant positive wealth effects of informal investigations of 0.20% (z-statistic = 1.37) over the (0, +1) event period, and 1.05% (z-statistic = 1.37) over the (0, +3) event period.

Formal investigations have statistically significant average wealth effects of −2.34% ($t = -5.94$) over the (−1, 0) event window. Nonparametric results for wealth effects tied to formal investigationannouncements are similarly significant at −1.17% (z-statistic = −3.13) over the (−1, 0) event window. Using both parametric and nonparametric methods, announcements of formal investigations exhibit negative wealth effects that are robust; significant negative wealth impacts are tied to formal investigation announcements for all six event-period windows.

Sample firms also realize smaller but still statistically significant average wealth effects of −1.19% ($t = -1.58$) on the announcement of a federal lawsuit filing over the (−1, 0) event period. Similarly significant average wealth effects of –1.43% ($t = -1.55$) are noted over the (−1, +1) event period, and −1.41% ($t = -1.88$) are noted over the (0, +1) event period. The robustness of such effects is demonstrated by statistically significant nonparametric wealth influences. Nonparametric results for wealth effects tied to suit filing announcements are similarly significant at −0.98% (z-statistic = −1.67) over the (−1, +1) event window, and −1.11% (z-statistic = −2.42) over the (0, +1) event window.

And finally, using parametric and nonparametric methods, there is scant evidence that firms who announce settlement agreements with the DOJ, FTC or SEC realize statistically significant wealth effects. The lone exception to this rule is found in median wealth effects of 0.74% ($z = 1.77$) over the (−1, +1) event period.

What results reported in Table 1 tell about the efficacy of federal law enforcement actions is ambiguous. In the case of informal investigations, formal investigations and lawsuits, negative abnormal returns suggest that the market regards such information as "bad news" for target firms. However, why such news is "bad" is unclear. There are at least three broad explanations for a negative response. First, announcements in *The Wall Street Journal* regarding enforcement activities might simply reveal information about illegal behavior not previously known by the market. Such information is bad because it represents news that the target cash flows will now be lower than expected. Second, firms whom the market did not believe would be caught are now the target of enforcement activity. That is, these results do not rule out the possibility that the market does have knowledge of illegal activities. Rather, these results may only indicate that the market is unable to precisely predict the probability of getting caught. And third, as suggested by Feroz, Park and Pastena (1991), effects of federal law enforcement actions on shareholder wealth may stem from the negative publicity generated and the possibility of third-party lawsuits.

Results reported in Table 1 also suggest that reports in *The Wall Street Journal* of a settlement agreement may represent "good news" concerning settlement negotiations not previously known to the market. Target firm cash flows may now be higher and more consistent than previously anticipated because the investigatory process was shorter than expected. Alternatively, following such announcements the market may be better informed regarding the (limited) scope of illegal activity by the target firm. Positive stock-price effects could also result if investors conclude that target firms are able to shield some aspect of their illegal advantage from enforcement agencies. Finally, positive abnormal returns could result if the market overestimates the probability or size of damage awards, fines, or other monetary penalties.

4.2 Announcement Effects by Violation Category

Table 2 summarizes market model cumulative average prediction error information by the actual or suspected violation category to provide additional perspective on (−1, 0) event-period wealth effects tied to the enforcement activities of the DOJ, FTC and SEC. Again, to test for robustness, both parametric and nonparametric results are reported.

Informal investigations delve into a wide variety of actual or suspected violations. The most common among these are 3 suspected accounting reporting violations, and 3 inside trading investigations. Based upon statistically significant parametric results, both of these common types of informal investigations give rise to meaningfully negative average wealth effects over the (−1, 0) event window. In the case of suspected accounting reporting violations, event-period average wealth effects of a whopping −18.07% ($t = -6.39$) are noted. Event-period average wealth effects of −9.25% ($t = -2.31$) are also large and statistically significant in the case of suspected inside trading activity. Less common informal investigation announcements tied to franchising practices and

Table 2. Cumulative Average Prediction Errors (CAPEs) by Actual or Suspected Violation Category.

Market-model cumulative average prediction errors (CAPEs), t-statistics, and z-statistics are shown for the (–1.0)-period windows for each type of actual or suspected violation category. Parametric and nonparametric statistics are shown, as are the size and statistical significance of mean (average) and median differences in announcement effects for various enforcement activities. The statistical significance of mean differences are evaluated using a t-test over the 180-day estimation period; median differences are evaluated using a z-test.

Violation Category	Sample Size	CAPEs	t–statistic	Median CAPEs	z–statistic
A. Informal Investigations (N = 13)					
Accounting reporting violations	3	−18.07%	−6.39 [c]	−19.40%	−1.76 [b]
Franchising practices	1	−6.30%	−1.51 [a]	−6.30%	−1.24
Fraudulent sales practices	1	3.66%	2.03 [b]	3.66%	1.02
Illegal foreign business practice	1	1.25%	0.68	1.25%	1.02
Illegal trading	1	1.64%	0.68	1.64%	1.00
Illegal use of company funds	1	0.58%	0.24	0.58%	0.90
Insider trading	3	−9.25%	−2.31 [b]	−14.09%	−0.41
Mergers and acquisitions	1	−1.77%	−0.57	−1.77%	−1.01
Violation of consent decree	1	5.44%	1.76 [b]	5.44%	0.98
B. Formal Investigations (N = 97)					
Accounting reporting violations	12	−10.33%	−5.72 [c]	−5.30%	−2.61 [c]
Attempting to monopolize	1	−0.97%	−0.52	−0.97%	−0.95
Bid rigging	1	−2.09%	−1.05	−2.09%	−0.90
Boycotts	3	1.21%	0.79	1.76%	0.64
Civil rights violations	1	−2.53%	−1.25	−2.53%	−1.01
Failure to report risk	1	−3.02%	−1.81 [b]	−3.02%	−1.00
Fraudulent sales practices	2	−3.93%	−2.65 [c]	−3.93%	−1.38 [a]
General restraints of trade	14	0.04%	0.06	0.82%	0.74
Illegal foreign business practice	1	1.49%	0.62	1.49%	0.99
Illegal trading	4	−28.47%	−8.53 [c]	−26.71%	−1.89 [b]
Illegal use of company funds	2	−2.16%	−0.37	−2.16%	0.07
Insider trading	25	2.84%	3.24 [c]	−1.12%	−1.79 [b]
Mergers and acquisitions	6	−2.05%	−2.31 [b]	−2.14%	−1.49 [a]
Price fixing	20	−1.26%	−1.86 [b]	−1.03%	−1.08
Questions regarding research results	2	−5.03%	−1.75 [b]	−5.03%	−1.35 [a]
Territorial divisions	1	1.02%	0.45	1.02%	1.13
Tying	1	0.95%	0.43	0.95%	1.01
C. Suits Filed (N = 28)					
Environmental violations	1	−4.57%	−1.32 [a]	−4.57%	−0.81
Failure to report risk	1	0.15%	0.09	0.15%	1.00
False advertising	2	6.23%	1.49 [a]	6.23%	1.24
Filing false claims with government	1	−6.04%	−0.72	−6.04%	−0.58
Fraudulent sales practices	1	−5.34%	−1.86 [b]	−5.34%	−0.97

Table 2. Continued.

Violation Category	Sample Size	CAPEs	t–statistic	Median CAPEs	z–statistic
Illegal trading	1	−3.30%	−1.12	−3.30%	−1.03
Insider trading	1	0.84%	0.20	0.84%	1.03
Mergers and acquisitions	9	−1.20%	−1.09	0.61%	0.40
Non-compliance with government production standards	1	−2.97%	−2.84 [c]	−2.97%	−0.90
Obstructing justice	1	−2.76%	−1.34 [a]	−2.76%	−1.14
Overcharging on government contracts	2	3.34%	3.10 [c]	3.34%	0.07
Price fixing	6	−1.98%	−1.34 [a]	−1.96%	−0.68
Questions regarding research results	1	−5.92%	−1.50 [a]	−5.92%	−0.89
D. Settlements (N = 24)					
Accounting reporting violations	7	4.95%	1.86 [b]	1.35%	0.78
Aiding fraudulent companies	1	5.13%	1.24	5.13%	1.12
Fraudulent sales practices	1	−2.22%	−0.87	−2.22%	−1.02
General restraints of trade	2	−1.15%	−0.63	−1.15%	−1.36 [a]
Illegal foreign business practice	1	−2.15%	−1.66 [b]	−2.15%	−0.95
Illegal trading	3	7.10%	1.34 [a]	1.80%	0.91
Mergers and acquisitions	1	0.77%	0.39	0.77%	1.05
Overcharging on government contracts	5	−1.71%	−1.74 [b]	−1.10%	−0.29
Poor certificate destruction	1	−0.49%	−0.15	−0.49%	−0.91
Unknown	1	−5.52%	−0.91	−5.52%	−0.91
Wire and mail fraud	1	−3.15%	−0.23	−3.15%	−0.72

[a] Significant at the 10% level (one-tail test).
[b] Significant at the 5% level (one-tail test).
[c] Significant at the 1% level (one-tail test).

violations of consent decrees also appear to engender statistically significant wealth effects, according to parametric findings. However, with the exception of suspected accounting reporting violations, where statistically significant event-period median wealth effects of −19.40% ($z = -1.76$) are noted, informal investigations do not generate median valuation effects that are statistically significant using nonparametric methods. Taken as a whole, these findings imply that the negative wealth effects tied to informal investigations are both somewhat smaller and less robust than the negative wealth effects associated with announcements concerning the launch of formal investigations. The lone exception to this rule may be informal investigations into suspected accounting reporting violations, where both parametric and nonparametric results suggest consistently large and statistically significant negative wealth effects.

As seen in Table 1, consistently large negative and statistically significant negative wealth effects are associated with announcements concerning the launch of formal investigations by the DOJ, FTC and SEC. As shown in Table 2, the most common types of formal investigations surround suspected accounting reporting violations (12), general restraints of trade (14), inside trading (25), mergers and acquisitions (6), and price fixing (20). Of these five violation categories, accounting reporting violations, mergers and acquisitions, and price fixing result in negative and statistically significant average wealth effects over the $(-1, 0)$ event window. In the case of accounting reporting violations, average wealth effects are -10.33% $(t = -5.72)$, and for mergers and acquisitions, average wealth effects are -2.05% $(t = -2.31)$. For price fixing, statistically significant average wealth effects of -1.26% $(t = -1.86)$ are noted. Formal investigation announcements tied to alleged inside trading results in a positive and statistically significant wealth effect of 2.84% $(t = 3.24)$. This anomalous result contrasts with a negative and statistically significant median reaction to investigations of inside trading of -1.12% $(z = -1.79)$. The 14 announcements of formal investigations initiated to look into accusations of general restraints of trade result in no significant wealth effects. Several less common formal investigations, including failure to report risk, fraudulent sales practices, illegal trading, and questions regarding research results, have statistically significant negative wealth effects. With the exception of the nonparametric results for inside trading, nonparametric statistics for formal investigation announcements are consistent with and corroborate parametric results.

Table 1 illustrates that average wealth effects associated with announcements concerning the filing of lawsuits are generally negative, but only weakly significant on a statistical basis. From Table 2, it seems clear that such negative influences stem from lawsuit announcements concerning alleged environmental violations $(-4.57\%, t = -1.32)$, fraudulent sales practices $(-5.34\%, t = -1.86)$, non-compliance with government production standards $(-2.97\%, t = -2.84)$, obstruction of justice $(-2.76\%, t = -1.34)$, price fixing $(-1.98\%, t = -1.34)$, and questions regarding research results $(-5.92\%, t = -1.50)$. Among these, lawsuit announcements concerning alleged price fixing tend to be most common. However, it is worth noting that when lawsuit filing announcements are sorted by violation category, none of the nonparametric test results show statistically significant median wealth effects. This suggests that the generally negative wealth effects tied to lawsuit filing announcements tend to be both smaller and less consistent than similar impacts of informal investigation and formal investigation announcements.

And finally, with respect to settlement announcements, Table 2 shows that average wealth effects are positive and statistically significant for settlements involving accounting reporting violations (4.95%, $t = 1.86$), and illegal trading (7.10%, $t = 1.34$). Conversely, average wealth effects are negative and statistically significant for settlements involving illegal foreign business practices (−2.15%, $t = -1.66$), and overcharging on government contracts (−1.71%, $t = 1.74$). The suggestion of a diverse market-value effect of settlement announcements is reenforced by the fact that median influences are negative and statistically significant only in the case of general restraint of trade settlements (−1.15%, $z = -1.36$). This diversity is a reasonable reflection of the fact that some settlement announcements can involve surprisingly large or punitive damages. Thus, it seems reasonable to conclude that the effect of settlement announcements on the value of the firm can be quite variable on a case-by-case basis.

4.3 Announcement Effects by Industrial Classification

Table 3 summarizes market model cumulative average prediction error information for the (−1, 0) event-period when such data are arrayed according to the industrial classification of the actual or suspected violator. Our purpose here is to learn if wealth effects tied to the enforcement activities of the DOJ, FTC and SEC vary systematically from one competitive environment to another, even though market structure can only be captured imperfectly at the very broad two-digit SIC industry sector level of aggregation. Chauvin and Hirschey (1993), for example, show that the market value of the firm is dependent, at least in part, upon intangible factors such as the magnitude of advertising and research and development activity. It seems plausible that the valuation effects of enforcement activities of the DOJ, FTC and SEC might be especially severe for companies that operate in a market environment where reputational capital, perhaps due to advertising, research and development, or other such factors, is especially important. Again, to test for robustness, both parametric and non-parametric results are reported.

Table 3 shows that negative and statistically significant average wealth effects related to informal investigations are most common for firms that make industrial and commercial machinery and computer equipment (−6.69%, $t = -3.32$), and manufacturers of electronic and other electrical equipment (−18.87%, $t = -1.90$). Companies that operate in these industries operate within some of the most R&D intensive sectors of the economy (see Chauvin & Hirschey, 1993). It stands to reason that the launch of informal investigations would have the potential for severe adverse effects on the reputational capital of such firms. An informal investigation announcement concerning a single miscellaneous retail establishment also had a negative and statistically significant wealth effect of a whopping −29.83% ($t = -6.11$). In this instance, the launch of an informal

Table 3. Cumulative Average Prediction Errors (CAPEs) by Industrial Classification of the Actual or Suspected Violator.

Market-model cumulative average prediction errors (CAPEs), t-statistics, and z-statistics are shown for the (–1.0)-period window according to the two-digit Standard Industrial Classification (SIC) Code of the actual or suspected violator. Parametric statistics and nonparametric statistics are shown, as are the size and statistical significance of mean (average) and median differences in announcement effects for various enforcement activities. The statistical significance of mean differences are evaluated using a t-test over the 180-day estimation period; median differences are evaluated using a z-test.

SIC Code	Industry Description	Sample Size	CAPEs	t-statistic CAPEs	Median	z–statistic
A. Informal Investigations (N = 13)						
20	Food and Kindred Products	1	0.58%	0.24	0.58%	0.90
35	Industrial and Commercial Machinery and Computer Equipment	4	−6.69%	−3.32 [c]	−6.42%	−0.02
36	Electronic and Other Electrical Equipment	1	−18.87%	−1.90 [b]	−18.87%	−0.76
48	Communications	1	−4.99%	−1.15	−4.99%	−0.93
53	General Merchandise Stores	1	−1.77%	−0.57	−1.77%	−1.01
59	Miscellaneous Retail	1	−29.83%	−6.11 [c]	−29.83%	−1.08
62	Security and Commodity Brokers, Dealers, Exchanges, and Services	2	2.65%	1.82 [b]	2.65%	1.43 [a]
73	Business Services	2	−0.55%	−0.19	−0.55%	−0.15
B. Formal Investigations (N = 97)						
2	Agricultural Production – Livestock and Animal Specialties	1	−2.14%	−0.34	−2.14%	−0.82
16	Heavy Construction – Nonbuilding Contractors	1	4.94%	2.36 [c]	4.94%	1.00
20	Food and Kindred Products	5	−2.61%	−2.05 [b]	−2.18%	−1.22
21	Tobacco Products	3	−1.04%	−0.74	−1.38%	−0.47
22	Textile Mill Products	1	−5.17%	−1.25	−5.17%	−1.06
25	Furniture and Fixtures	1	−6.71%	−3.21 [c]	−6.71%	−1.00
27	Printing, Publishing, and Allied Industries	5	−0.59%	−0.47	−0.58%	−2.07 [b]
28	Chemicals and Allied Products	6	−1.93%	−2.17 [b]	−1.82%	−1.51 [a]
30	Rubber and Miscellaneous Plastic Products	2	−2.58%	−1.46 [a]	−2.58%	−1.41 [a]
32	Stone, Clay, Glass, and Concrete Products	3	−1.64%	−0.88	−2.18%	−0.49
33	Primary Metal Industries	4	0.49%	0.27	0.83%	1.12
35	Industrial and Commercial Machinery and Computer Equipment	6	−4.32%	−2.32 [b]	−2.32%	−0.53

Table 3. Continued.

SIC Code	Industry Description	Sample Size	CAPEs	t-statistic CAPEs	Median	z-statistic
36	Electronic and Other Electrical Equipment	12	−3.73%	−3.68 [c]	−1.03%	−0.92
37	Transportation Equipment	7	−0.77%	−0.51	−0.61%	−0.29
38	Measuring, Analyzing, and Controlling Instruments; Photographic, Medical and Optical Goods; Watches and Clocks	4	−10.25%	−4.24 [c]	−4.62%	−1.87 [b]
39	Miscellaneous Manufacturing Industries	1	−10.52%	−3.31 [c]	−10.52%	−0.98
41	Local and Suburban Transit and Interurban Highway Passenger Transportation	1	−6.19%	−0.84	−6.19%	−1.06
45	Transportation by Air	2	2.22%	1.27	2.22%	1.44 [a]
48	Communications	5	−3.92%	−1.21	−2.89%	−0.20
49	Electric, Gas, and Sanitary Services	1	−4.71%	−0.70	−4.71%	−0.67
51	Wholesale Trade − Nondurable Goods	1	−73.10%	−10.07 [c]	−73.10%	−0.96
53	General Merchandise Stores	1	−2.41%	−1.08	−2.41%	−1.06
59	Miscellaneous Retail	1	1.02%	0.45	1.02%	1.13
60	Depository Institutions	2	−1.95%	−1.60 [a]	−1.96%	−1.36 [a]
62	Security and Commodity Brokers, Dealers, Exchanges, and Services	2	−0.16%	−0.11	−0.16%	0.09
67	Holding and Other Investment Offices Services	8	−0.85%	−0.77	−0.36%	−0.48
73	Business Services	8	9.62%	5.94 [c]	−0.56%	−0.56
80	Health Services	2	−12.65%	−2.36 [c]	−12.65%	−1.32 [a]
95	Environmental Services	1	7.45%	2.93 [c]	7.45%	1.03
C. Suits Filed (N = 28)						
10	Metal Mining	1	−0.64%	−0.24	−0.64%	−1.01
21	Tobacco Products	1	2.94%	1.09	2.94%	1.05
22	Textile Mill Products	1	−4.57%	−1.32 [a]	−4.57%	−0.81
25	Furniture and Fixtures	1	−2.76%	−1.34 [a]	−2.76%	−1.14
31	Leather and Leather Products	1	7.18%	0.86	7.18%	0.73
34	Fabricated Metal Products (nontransportation, nonmachinery)	1	0.61%	0.36	0.61%	0.94
35	Industrial and Commercial Machinery and Computer Equipment	3	−3.84%	−2.27 [b]	−4.86%	−0.58
36	Electronic and Other Electrical Equipment	3	−0.73%	−0.52	−2.97%	−0.41
37	Transportation Equipment	1	−0.01%	−0.01	−0.01%	−0.99

Table 3. Continued.

SIC Code	Industry Description	Sample Size	CAPEs	t-statistic CAPEs	Median	z–statistic
38	Measuring, Analyzing, and Controlling Instruments; Photographic, Medical and Optical Goods; Watches and Clocks	2	3.06%	1.26	3.06%	1.47 [a]
45	Transportation by Air	6	−1.98%	−1.34 [a]	−1.96%	−0.68
50	Wholesale Trade – Durable Goods	1	−6.04%	−0.72	−6.04%	−0.58
60	Depository Institutions	1	0.15%	0.09	0.15%	1.00
62	Security and Commodity Brokers, Dealers, Exchanges, and Services	1	−3.30%	−1.12	−3.30%	−1.03
67	Holding and Other Investment Offices Services	1	−5.34%	−1.86 [b]	−5.34%	−0.97
73	Business Services	3	−0.73%	−0.28	1.36%	0.71
D. Settlements (N = 24)						
27	Printing, Publishing, and Allied Industries	2	2.80%	0.62	2.80%	0.08
28	Chemicals and Allied Products	2	1.06%	0.47	1.06%	1.64 [a]
35	Industrial and Commercial Machinery and Computer Equipment	1	−0.17%	−0.05	−0.17%	−0.97
36	Electronic and Other Electrical Equipment	4	−3.81%	−2.27 [b]	−3.84%	−0.90
37	Transportation Equipment	1	−1.10%	−0.52	−1.10%	−0.98
38	Measuring, Analyzing, and Controlling Instruments; Photographic, Medical and Optical Goods; Watches and Clocks	3	14.26%	2.02 [b]	21.85%	1.09
51	Wholesale Trade – Nondurable Goods	1	−0.02%	−0.01	−0.02%	−0.67
63	Insurance Carriers	2	1.17%	0.33	1.17%	0.08
67	Holding and Other Investment Offices Services	5	0.42%	0.33	−0.49%	−0.35
73	Business Services	2	0.03%	0.03	0.03%	0.14
80	Health Services	1	−1.03%	−0.07	−1.03%	−0.81

[a] Significant at the 10% level (one-tail test).
[b] Significant at the 5% level (one-tail test).
[c] Significant at the 1% level (one-tail test).

investigation might be expected to engender the risk of a significant loss in reputational capital stemming from advertising expenditures. In a somewhat puzzling manner, positive and statistically significant average wealth effects are tied to informal investigation announcements concerning two security and commodity brokers (2.65%, $t = 1.82$). Using nonparametric methods, the median wealth effect of 2.65% ($z = 1.43$) is also significant. At least in these two instances, investors may have been relieved that more serious enforcement activity, such as a formal investigation, was not announced.

Table 3 also shows that consistently negative average wealth effects associated with formal investigation announcements are most common for firms operating in food and kindred products (−2.61%, $t = -2.05$), chemicals and allied products (−1.93%, $t = -2.17$), rubber (−2.58%, $t = -1.46$), industrial and commercial machinery and computer equipment (−4.32%, $t = -2.32$), electronic and other electrical equipment (−3.73%, $t = -3.68$), and measuring instruments (−10.25%, $t = -4.24$). These generally adverse average wealth effects tied to formal investigation announcements tend to be corroborated by nonparametric results concerning median effects. Interestingly, with the exception of food and kindred products, where advertising is very important, each of the other industries mentioned above is highly R&D intensive. This lends support to arguments that negative wealth effects tied to the launch of formal investigations by the DOJ, FTC, or SEC are most severe for firms that stand to lose significant amounts of reputational capital derived from intangible assets, like advertising and R&D.

And finally, Table 3 also shows the industrial affiliations of companies with typically negative average wealth effects associated with lawsuit filing announcements, and the mixed average wealth effects generated by settlement announcements. Consistently negative average wealth effects associated with lawsuit filing announcements are most common for firms operating in industrial and commercial machinery and computer equipment (−3.84%, $t = -2.27$), and air transportation (−1.98%, $t = -1.34$). Settlement announcements commonly result in negative average wealth effects for firms that produce electronic and other electrical equipment (−3.81%, $t = -2.27$), but positive average wealth effects for firms that produce measuring instruments (14.26%, $t = 2.02$). However, all such effects appear to be less consistent than wealth effects tied to both informal and formal investigation announcements in that they do not tend to be corroborated by nonparametric results concerning median effects. Thus, one must be somewhat more cautious in arguing that negative wealth effects associated with lawsuit and/or settlement announcements by the DOJ, FTC, or SEC are most severe for firms that stand to lose significant amounts of reputational capital.

5. SUMMARY AND CONCLUSIONS

This study provides the first large-sample analysis of the stock-market reaction to public announcements concerning the federal law enforcement activities of the DOJ, FTC, and SEC. Evidence of wealth effects are presented for a variety of enforcement activities including: informal investigations, formal investigations, lawsuits, and settlement agreements. Moreover, an analysis is provided of average wealth effects by the category of actual or suspected violation, and by the industrial classification of the actual or suspected violator. Our purpose here is to learn more about economic characteristics that might drive wealth effects tied to the enforcement activities of the DOJ, FTC and SEC.

In terms of results, we find that negative abnormal returns are consistently observed over event periods in which news of a DOJ, FTC and SEC investigations (both informal and formal) or lawsuits appear in *The Wall Street Journal*. Evidence on the average wealth effects tied to settlement announcements is mixed, with both negative and positive abnormal returns evident on a case-by-case basis. Negative abnormal returns are observed for news regarding settlements involving illegal foreign business practices and overcharging on government contracts; positive abnormal returns are noted for settlements of accounting reporting violations and illegal trading activity. Both types of results suggest that prior to *The Wall Street Journal* announcement date these government bodies posses information that is relevant to the determination of future cash flows for target firms. Because such announcements consistently result in statistically significant stock-price effects, they can be regarded as important in the formation of investor expectations concerning future cash flows.

Findings reported here have direct relevance for competitive policy in dynamic markets, and for the corporate governance literature. Our findings lend empirical support to arguments that negative wealth effects tied to the launch of enforcement activity by the DOJ, FTC, or SEC are most severe for firms that stand to lose significant amounts of reputational capital. Moreover, analysis of the competitive environment of affected firms suggests that such reputational capital is derived from intangible assets, like advertising and R&D. Federal law enforcement activities of the DOJ, FTC and SEC may thus be seen as part of what Jensen and Meckling describe as monitoring mechanisms inside and outside the firm that work together to establish an optimal set of restrictions on firm activity, including the investment in and management of intangible assets. With this context, public announcements regarding federal law enforcement actions represent a flexible tool that can be used to facilitate competition policy in dynamic markets.

NOTES

1. For 54 of the 58 firms in the Feroz, Park and Pastena (1991) sample, day (0) is the announcement date in *The Wall Street Journal*, or in a publication followed in the *Funk and Scott's Index*. In the other four instances, day (0) is the registrant's press release disclosure date.

2. A few additional miscellaneous announcements of enforcement activities of the DOJ, FTC and SEC were also noted including: 9 Civil Investigative Demands (CIDs), 4 simultaneous announcements of a suit being filed and settled, 3 Wells Submissions, 5 administrative actions, 3 convictions and/or guilty pleas, and 8 announcements of investigations being ended. A detailed analysis revealed no statistically significant valuation effects of each of these types of announcements. Estimation results are available on request.

3. Cumulative average prediction errors were also estimated using the market model and an equally-weighted index, the market adjusted return model and a value-weighted index, the market adjusted return model and an equally-weighted index, the standardized abnormal return market model and an equally-weighted index, and the standardized abnormal return market model and a value-weighted index. These estimation results are substantially the same and available on request.

4. An exploratory cross-sectional regression analysis was also conducted to learn if any consistent relationship is present between event-period CAPEs and the ownership structure of the firm, various firm size measures, or indicators of firm performance and leverage. No such substantive linkages were discovered.

REFERENCES

Bhagat, S., Brickley, J. A., & Coles, J. L. (1994). The Costs of Inefficient Bargaining and Financial Distress. *Journal of Financial Economics*, *35* (April), 221–247.

Butler, H. N. (1987). *Legal Environment of Business*. South-Western Publishing Co., Cincinnati, Ohio.

Bittlingmayer, G. (1993). The Stock Market and Early Antitrust Enforcement. *Journal of Law and Economics*, *36* (April), 1–31.

Chauvin, K. W., & Hirschey, M. (1993). Advertising, R&D Expenditures and the Market Value of the Firm. *Financial Management*, *22* (Winter), 128–140.

Corrado, C. J. (1989). A Nonparametric Test for Abnormal Security-Price Performance in Event Studies. *Journal of Financial Economics*, *23* (August), 385–395.

Fama, E. F., & Jensen, M. C. (1983). Separation of Ownership and Control. *Journal of Law and Economics*, *26* (June), 301–325.

Feroz, E. H., Park, K., & Pastena, V. S. (1991). The Financial and Market Effects of the SEC's Accounting and Auditing Enforcement Releases. *Journal of Accounting Research*, *29* (Supplement), 107–142.

Garbade, K. D., Silber, W. L., & White, L. J. (1982). Market Reaction to the Filing of Antitrust Suits: An Aggregate and Cross-Sectional Analysis. *The Review of Economics and Statistics*, *64* (November), 686–691.

James, C. (1987). Some Evidence on the Uniqueness of Bank Loans. *Journal of Financial Economics*, *19* (December), 217–235.

Jensen, M. C., & Meckling, W. H. (1976). Theory of the Firm: Managerial Behavior, Agency Costs and Ownership Structure. *Journal of Financial Economics*, *3* (October), 305–360.

Myers, S. C. (1977). Determinants of Corporate Borrowing. *Journal of Financial Economics*, *5* (November), 147–175.

EMPIRICAL EVIDENCE ON DETERMINANTS OF CAPITAL STRUCTURE

Tomas Jandik and Anil K. Makhija

ABSTRACT

Based on received financial theory, we empirically examine the role of the following firm-specific determinants of leverage: bankruptcy costs, growth, variability, non-debt tax shields, collateral value, profitability, and size. For our sample, to focus on firm-specific aspects, we purposely use pooled time-series cross-sectional data from a single industry (electric and gas utilities) for the twenty-year period, 1975–1994. Our findings largely support theory, with the important exception of variability. We find that leverage is positively related with variability, contrary to the literature. We conjecture that this seemingly perverse relationship may be due to the yet unrecognized effect of variability: firms with greater variability of earnings have a greater chance that their non-debt tax shields may prove to be inadequate, and are therefore expected to take on higher levels of leverage.

1. INTRODUCTION

Harris and Raviv (1991) provide a comprehensive review of the determinants of capital structure, stressing largely cross-sectional differences. We examine the determinants of capital structure by examining cross-sectional and inter-temporal changes. In particular, we consider the case of electric utilities because they

Advances in Financial Economics, Volume 6, pages 143–159.
Copyright © 2001 by Elsevier Science B.V.
All rights of reproduction in any form reserved.
ISBN: 0-7623-0713-7

experienced dramatic changes in their total debt to assets and long-term debt to assets ratios, with both ratios falling by a remarkable 10% figure over the twenty-year period, 1975–1994. The study of capital structure changes within a single industry has certain advantages. Those firm characteristics that reflect the production technology of firms, such as operating leverage, are largely automatically controlled for in a study of firms within a single industry. Similarly, macro-economic factors are expected to impact firms in similar ways within an industry. Thus, we are able to focus on firm-specific determinants of capital structure, compared with the typical study that investigates a sample consisting of a broader cross-section of firms drawn from many industries. We also purposely do not emphasize differential regulatory effects in this study, since we are primarily interested in financial theory and the determinants of capital structure that it predicts. Admittedly, regulatory factors may have a significant role to play here. We, however, are interested only in the explanatory power of received financial theory.

We examine the role of several factors that have been identified in prior finance literature as the major determinants of capital structure. Specifically, we investigate bankruptcy costs, growth, variability, non-debt tax shields, collateral value, profitability, and size. There is an extensive literature supporting these factors, with sometimes conflicting empirical evidence.

Our empirical design is straightforward. We form a sample of electric and gas utilities (SIC codes 4911 and 4931) for the period 1975–1994. Then, in a pooled cross-sectional time-series regression, we explain their capital structure in terms of our set of chosen determinants.

In general, our findings support the predicted role of the determinants of leverage proposed in the previous literature with the important exception of variability (risk). While the direct effect of risk, through bankruptcy costs and other factors, would suggest a negative relation between leverage and risk, our empirical evidence suggests a positive relation. We explain this as a yet unrecognized indirect effect of risk. Other things equal, risk lowers the value of non-debt tax shields. Since the level of leverage is inversely related to non-debt tax shields, we may observe the seemingly perverse positive relation between risk and leverage.[1]

The rest of the paper is organized as follows. In the next section, we review the literature and formulate hypotheses regarding the determinants of capital structure. In Section 3, we describe the data, present our regression methodology, and discuss the empirical findings. Our conclusions are in Section 4.

2. HYPOTHESES: DETERMINANTS OF LEVERAGE

In this section, we describe the major determinants of capital structure identified in previous literature. In each case we cite relevant research, and state the predicted effect on capital structure.

2.1. Bankruptcy Costs (Diversification)

Altman (1984), Ang, Chua, and McConnell (1984), and Warner (1977) provide evidence on direct and indirect costs brought on by bankruptcy. Greater expected bankruptcy costs result as a consequence of increases in leverage. The problem, however, is how to measure and empirically incorporate potentially higher bankruptcy costs associated with greater leverage. It is, therefore, not surprising that the existing empirical literature has not tested the role of bankruptcy costs directly. Instead, prior research points at other firm characteristics that are likely to be associated with bankruptcy costs, such as return variability. While variability in equity returns may play a role in this context, it is perhaps not the best variable to proxy bankruptcy costs when studying the changes in leverage. After all, according to Modigliani and Miller (1958, 1963), higher leverage itself produces greater variability of equity returns. In this study, we proxy bankruptcy costs by a measure of the inverse of diversification. We expect that less diversified firms are more likely to experience earnings shocks that result in insolvency.

2.2. Growth

Kim and Sorensen (1986), Kester (1986), Titman and Wessels (1988), and Chaplinsky and Niehaus (1990) all consider growth as a determinant of capital structure, with conflicting evidence. The theory suggests a negative relationship. Unlike assets in place, the value of growth opportunities is subject to severe agency problems. Jensen and Meckling (1976), Myers (1977), and Smith and Watts (1992) point out that by risk-shifting or by sub-optimal investment, stockholders can change investment policy (negatively affecting company's value and risk characteristics) as well as growth opportunities. Thus, lenders are likely to charge higher interest rates when a greater proportion of the firm's value is derived from future growth opportunities.

Following the previous work in this area, we proxy growth with Stock Market Price to Book Value of Equity ratio. We predict a negative relation between this market-to-book ratio and leverage.

2.3. Non-Debt Tax Shields

DeAngelo and Masulis (1980) argue that firms that have alternatives to protect their earnings from taxation will need less debt (because of a lesser need to deduct interest expenses in order to lower taxes). Thus, we expect non-debt tax shields to be negatively related to leverage. The empirical evidence on this relationship is, however, mixed. Bowen, Delay, and Huber (1982) found a negative relationship between non-debt tax shields and the debt ratios, as predicted by DeAngelo and Masulis (1980). On the other hand, the findings of Boquist and Moore (1984) and Bradley, Jarrell, and Kim (1984) failed to support a negative relationship.

We employ three different measures of non-debt tax shields: Depreciation/Total Assets, Imputed Non-debt Tax Shield/Total Assets, and Investment Tax Credits/Total Assets. Depreciation and Investment Tax Credits (which were applicable for a considerable time period covered by our study) are obvious substitutes for the tax benefits of the deductibility of interest on debt. These non-debt tax shields have also been used by Bradley, Jarrell, and Kim (1984) and Titman and Wessels (1988). In addition, we also compute imputed non-debt tax shields as follows: Since

Federal Taxes = Corporate Tax Rate* (Operating Income – Interest Payments – Non-debt Tax Shields),

we can compute:

Non-debt Tax Shields = Operating Income − Interest Payments – (Federal Taxes/Corporate Tax Rate)

2.4. Variability

Recent studies such as Rajan and Zingales (1995) have ignored this factor, perhaps because of problems with the availability of data. Earlier research, however, takes this variable into consideration. Kim and Sorensen (1986) used the coefficient of variation in earnings. Kester (1986) used the variance of residuals from a regression (explaining adjusted earnings). Titman and Wessels (1988) used the standard deviation of changes in operating income, and Bradley, Jarrell and Kim (1984) utilized the standard deviation of first differences in earnings. In each of the cases, the researchers tried to form measures that are not expected to be influenced by capital structure itself (in the manner described by Modigliani-Miller Theorems, which suggest that leverage increases stock price volatility). Rather, they tried to capture the variability of assets returns.

On the theoretical side, one can identify a number of reasons why variability should affect leverage. Higher variability is suggestive of a more uncertain environment in which observability of managerial actions is limited and debt providers fear greater agency-related expropriation by equity holders. Greater variability of earnings may also lead to greater probability of bankruptcy and, consequently, to an increase in expected bankruptcy costs (Warner, 1977; Scott,

1977; Weiss, 1990). Thus, we predict lower leverage for firms with greater variability of earnings.

We use standard deviation of the percentage change in operating income as our measure of variability. (We do not use equity-based variables such as beta, since such measures would be inappropriate, again according to Miller and Modigliani Theorems).

2.5. Collateral Value

The presence of collateralizable debt reduces the agency costs, and ultimately increases the ability of a firm to issue more debt. In addition, more collateralizable assets are associated with smaller proportion of growth options with respect to firm value.

We follow Titman and Wessels (1988) and use Inventory, Gross Plant and Equipment divided by Total Assets as a proxy of collateral value. In addition, we also use Inventory/Total Assets.

2.6. Profitability

According to their Pecking Order Hypothesis, Myers and Majluf (1984) argue that firms prefer to finance their capital expenditures using retained earnings rather than external financing such as debt. More profitable firms have more retained earnings and thus are expected to use lower leverage. Consequently, we predict a negative relation between leverage and our measure of profitability (defined as Operating Income/Total Assets).

2.7. Size

Reporting a concave relationship between direct bankruptcy costs and the size of the firm, Ang, Chua and McConnell (1982), as well as Warner (1977) suggest a scale effect between size and losses due to bankruptcy. Since we already take into consideration the effect of bankruptcy costs, we also include a size measure in our analysis. In addition, the expected costs of financial distress are likely to be lower for large (arguably older, more stable) firms according to Warner (1977), and Weiss (1990). Lenders should be willing to provide more debt to such firms. Lastly, debt may be the most effective mechanism to eliminate free cash flow problems (Jensen, 1986) in large firms, where implementation of alternative governance mechanisms (blockholdings, market for corporate control) may be relatively costly. As a result, we predict a positive relation between leverage and size, with size defined as the logarithm of Total Sales.

Table 1. Time Series.

Leverage, growth, non-debt tax shield, variability of earnings, and effect of diversification of electric and gas utilities: yearly averages for firms in SIC codes 4911 and 4931, 1975–1994.[1, 2]

	Leverage		Growth	Non-debt tax shields			Earnings variability	Diversification
Year	TD/TA	LTD/TA	MVE/ BVE	NDT/TA $*10^{-2}$	ITC/TA $*10^{-4}$	ITC*/TA $*10^{-4}$	SIGMA	AHERF
1975	0.48	0.43	0.91	6.11	7.63	7.63	–	–
1976	0.46	0.43	1.04	6.63	10.96	10.96	–	–
1977	0.45	0.42	1.00	6.69	12.59	12.59	–	–
1978	0.44	0.40	0.87	6.29	12.87	12.87	–	–
1979	0.43	0.39	0.78	6.03	12.81	12.81	0.1104	–
1980	0.43	0.39	0.73	5.67	13.53	13.53	0.1260	0.8959
1981	0.42	0.38	0.77	5.75	15.40	15.40	0.1373	0.8956
1982	0.41	0.37	0.93	6.34	14.66	14.53	0.1409	0.8919
1983	0.39	0.36	0.94	6.66	15.04	15.36	0.1449	0.8857
1984	0.39	0.37	1.03	6.51	17.60	17.26	0.1366	0.8831
1985	0.39	0.36	1.22	6.42	16.43	17.33	0.1270	0.8729
1986	0.39	0.39	1.44	6.20	18.56	18.56	0.1189	0.8709
1987	0.39	0.35	1.18	5.58	19.30	17.64	0.1122	0.8629
1988	0.39	0.34	1.25	4.66	12.92	18.74	0.1063	0.8424
1989	0.38	0.34	1.42	4.81	17.00	16.45	0.1025	0.8399
1990	0.39	0.34	1.30	4.11	14.44	8.38	0.0972	0.8323
1991	0.40	0.35	1.57	4.33	14.56	8.77	0.0974	0.8280
1992	0.39	0.35	1.58	4.83	13.62	8.14	0.1042	0.8244
1993	0.37	0.33	1.59	4.67	12.77	7.28	0.1060	–
1994	0.37	0.33	1.30	4.44	11.28	5.82	0.1098	–

[1] TD/TA = Total Liabilities / Total Assets
LTD/TA = Long-term Debt / Total Assets
MVE/BVE = Market-to-Book Ratio of Equity
NDT/TA = Imputed Non-debt Tax Shield / Total Assets
ITC/TA = Investment Tax Credit/Total Assets
SIGMA = Standard Deviation of Percentage Changes in Operating Income over past 5 years incl. the year of interest
AHERF = Asset Herfindahl Index.

[2] See Table 2 for sources of data and further details on the variables.

3. DATA, METHODOLOGY, AND ANALYSIS

3.1. A Time Series Analysis

We begin with a time series examination of leverage of electric utilities in SIC codes 4911 and 4931 for the 20-year period (1975–1984). The trend in debt use is shown in Table 1. Both measures of leverage, Total Debt-to-Assets

(TD/TA) and Long-term Debt-to-Assets (LTD/TA) decline over the period. In each case, the average falls by about 10%. With this remarkable change in capital structure, we have a feasible empirical experiment for an examination of the role of various determinants of leverage proposed in the literature. Turning to selected determinants that have been highlighted before, we report growth (MVE/BVE), non-debt tax shield (3 different measures), and the variability of earnings (SIGMA). In addition, we document the evolution of an Asset Herfindahl Index (AHERF), an inverse measure of diversification.

The decline in leverage is accompanied by changes in the determinants of leverage according to Table 1. As predicted by the hypothesized negative relation between leverage and growth, MVE/BVE increased noticeably over the sample period. Non-debt tax shields (NDT/TA), however, do not follow the hypothesized pattern. This measure drops over time, whereas we expected it to increase (going counter to the observed decline in leverage over the sample period). There is also some decline in earnings variability across time. This change, however, is not statistically significant.

Table 1 also shows that diversification of sample firms increased over time, as evidenced by the decline of asset Herfindahl Index (AHERF). This is an expected change – leverage and diversification move in opposite directions.

3.2. Proxies for leverage and its determinants

Throughout the remaining analysis, we consider four alternative measures of leverage. The measures based on book values are:

TD/TA = Total Liabilities/Total Assets
LTD/TA = Long-term Debt/Total Assets

In addition, we used the following two measures utilizing market values:

TD/MVA = Total Liabilities/Imputed Market Values of Assets
LTD/MVA = Long-term Debt/Imputed Market Value of Assets

The imputed market value of assets is calculated as the market value of equity plus book value of debt and preferred stocks (since market values of debt and preferred stocks are not readily reported). We expect managers to pay greater attention to market values, although there is no significant empirical evidence of such managerial behavior in this regard.

The definitions of the determinants of leverage were provided alongside the above hypotheses. For convenience, all the variables used in the following analysis are listed in Table 2, including the pertinent Compustat definitions and other details such as the hypothesized relation with leverage.

Table 2. Proxies for Leverage and Its Determinants: Definitions and Hypothesized Relations.

Panel A describes four measures of leverage that are employed as dependent variables in the subsequent analysis. Panel B contains a description of the independent variables. Firm level accounting data are obtained from Compustat Industrial files. Line of business data are taken from Compustat Industrial Segment files. Stock returns data are obtained from CRSP (Center for Research in Security Prices) tapes, which are available from the University of Chicago

Panel A: Proxies for leverage

Variable	Definition	Compustat Variables Used
TD/TA	Total Liabilities/Total Assets	(#9+#34)/(#6)
LTD/TA	Long-term Debt/Total Assets	(#9)/(#6)
TD/MVA	Total Liabilities/Imputed Market Asset Value	(#9+#34)/ (#24*#25+#9+#34+#130)
LTD/MVA	Long-term Debt/Imputed Market Asset Value	(#9)/ (#24*#25+#9+#34+#130)

Imputed Market Value of Assets = Market Value of Equity plus Book Value of Debt.

Panel B: Determinants of Leverage

Variable	Hypothesized Sign of Coef.	Definition	Compustat Variables Used and Other Details
INV/TA	+	Inventory / Total Assets	(#3)/(#6)
MVE/BVE	–	Market-to-Book Ratio of Equity	(#24*#25)/(#60)
AHERF	+	Asset Herfindahl Index of assets devoted to	Sum of squares of fractions for different lines of business
IGP/TA	+	Inventory, Gross Plant & Equipment/Total Assets	(#3+#7)/(#6)
OI/TA	–	Operating Income/ Total Assets	(#13)/(#6)
LSALES	+	Logarithm of Sales	Log(#12)
DPR/TA	–	Depreciation/Total Assets	(#14)/(#6)
NDT/TA	–	Non-debt Tax Shield/ Total Assets	[Operating Income – Interest Expense – (Federal Taxes/ Corporate Tax Rate)] /(Total Assets) The corporate tax rate was 48% from 1975 to 1978, 46% from 1979 to 1986, 42.5% in 1987, and 34% thereafter.
ITC/TA	–	Investment Tax Credit/Total Assets	(#51)/(#6)
SIGMA	+	Standard deviation of percentage changes in Operating Income	Percentage changes in (#13) for the past five years, including the year of interest are estimated. Then, the standard deviation of those percentages is taken.

3.3. Descriptive Statistics

Table 3 provides the mean, median, and standard deviation for each of the variables used in the analysis. We pool the data for all the years covered, assuming that there is a stable, time-invariant relationship between leverage and its determinants. (As a check, we considered two sub-periods as well as the full period in our analysis. Only the full period results are reported, since the analysis of sub-periods did not yield significant differences).

Table 3 shows that except for three variables, the means and medians are generally similar, suggesting that there are few outliers in our sample. For IGP/TA and SALES, sample means and medians differ noticeably. The difference for SALES is not surprising, since there are some very large firms in our sample. The collateral values of certain firms appear to be large, since the mean of IGP/TA (1.45) is considerably greater than the median (1.15). Interestingly, this difference is apparently not due to inventories, since the mean and median of INV/TA are not statistically different from each other.

It should also be noted that the values of standard deviations (with respect to the mean) of the determinants of leverage suggest that the values of the determinants vary considerably across sample firms. This suggests that regression analysis is an appropriate tool to determine the impact of factors hypothesized to affect leverage choices.

3.4. Regression Analysis of Leverage

Table 4 presents the regression analysis of leverage using pooled time-series cross-sectional data for electric and gas utilities with SIC codes 4911 and 4931 for the period 1980–1992. Data for earnings variability (SIGMA) is not available till 1980 because we use a five-year window to form the standard deviation of percentage changes in operating income. The last year of the analysis is 1992 because of the unavailability of diversification (AHERF) data. The analysis in Table 4 is presented in three panels as we consider three alternative definitions of non-debt tax shields.

In panel A of Table 4, we present findings for Depreciation / Total Assets (DPR/TA), as the measure of non-debt tax shields. The adjusted R^2 for regressions using book measures of debt are 29.60% and 29.00%. The corresponding adjusted R^2 for regressions with market measures of leverage are much higher, 66.36% and 61.58%, consistent with the assumption that managers should be more concerned with market values. However, in estimation of both book and market leverages, most of the proposed determinants appear to influence firm capital structure choices in a manner consistent with the theory.

Table 3. Descriptive Statistics for Leverage and Its Determinants for Electric and Gas Utilities, 1975-1994.

End of year data from Compustat tapes are used to compute statistics for 134 electric and gas utilities in SIC codes 4911 and 4931 for the period 1975-1994. Stock returns are obtained from CRSP tapes. Data for AHERF are obtained from the Industry Segment files compiled by Compustat.

Variable	No. of Obs.	Mean	Median	Std. Dev.	Variable	No.of Obs.	Mean	Median	Std. Dev.
TD/TA	2421	0.4067	0.4045	0.0834	IGP/TA	2415	1.4456	1.1459	0.1599
LTD/TA	2421	0.3676	0.3692	0.0801	OI/TA	2413	0.1178	0.1170	0.0270
TD/MVA	2128	0.5040	0.5108	0.0925	LSALES	2422	6.3847	6.4912	1.3659
LTD/MVA	2128	0.4558	0.4637	0.0864	DPR/TA	2418	0.0309	0.0303	0.0090
MVE/BVE	2261	1.1270	1.0540	0.4249	NDT/TA	2080	0.0560	0.0576	0.0235
INV/TA	2415	0.0315	0.0292	0.0150	ITC/TA	1797	0.0015	0.0012	0.0013
AHERF	1349	0.8629	1.0000	0.1822					

TD/TA = Total Liabilities / Total Assets
LTD/TA = Long-term Debt / Total Assets
TD/MVA = Total Liabilities / Imputed Market Value of Assets
TD/MVA = Long-term Debt / Imputed Market Value of Assets
MVE/BVE = Market-to-Book Ratio of Equity
INV/TA = Inventory / Total Assets
AHERF = Asset Herfindahl Index

IGP/TA = Inventory, Gross Plant, and Equipment / Total Assets
OI/TA = Operating Income / Total Assets
LSALES = Logarithm of Sales
DPR/TA = Depreciation / Total Assets
NDT/TA = Non-debt Tax Shield / Total Assets
ITC/TA = Investment Tax Credit / Total Assets

See Table 2 for additional details, including variable numbers according to Compustat

MVE/BVE has a significantly negative coefficient in every estimation reported in Table 4, affirming the hypothesized negative relation between growth and leverage. Profitability and leverage also appears to have a significant negative relation, since the coefficient of OI/TA is negative in all estimations and it is significantly different from zero at the 1% level in every case. When we consider inventory as a proxy of collateral value, we observe the predicted positive sign for the coefficient of INV/TA in all estimations. Except for two specifications, the coefficient is statistically significant.

Generally, the findings also suggest that larger firms, as expected, have greater leverage. This is clearly the case in panel B where the reported coefficients of LSALES are all positive and significant at the 1% level in three out of four cases. Also in panel C, when significant, the coefficients of LSALES are positive. The only conflicting evidence regarding the relation between size and leverage seems to be associated with results of panel A, but the relevant regressions involve book, not market, measures of leverage.

The hypothesized inverse relation between leverage and non-debt tax shields is supported by the results of our analysis. In panel C, where we use ITC/TA as the non-debt tax shields proxy, the regression coefficient is negative and statistically significant in all four estimations. In panel A, the findings support the hypothesized negative relation in two cases. However, the one case where the results do not correspond with prediction, involves a market measure of leverage. In panel B, none of the findings are significant, possibly because our proxy – imputed value of non-debt tax shields – suffers from errors-in-variable problems. If our empirical measure is too noisy a proxy of actual non-debt tax shields, the insignificant findings may be expected.

Some of the findings in Table 4 contradict the hypothesized relationship between leverage and its determinants. While the collateral value of inventories was observed to have a positive effect on leverage, the coefficient of IGP/TA is negative whenever it is significant in various regressions reported in Table 4. These findings are unexpected, but they may be a consequence of special difficulties faced by capital-intensive firms in this industry. Still, the other measure of collateral (INV/TA), which is less likely to be related to the degree of capital intensiveness of firms in our sample, yields the expected coefficients.

Another variable with unexpected findings is AHERF, a measure of diversification. Less diversified firms have higher AHERF values. Since we expect these firms to be riskier and bankruptcy-prone, they should carry smaller proportions of debt in their capital structures. However, the evidence in Table 4 strongly suggests otherwise. Except for one estimation, AHERF has positive coefficients at 1% significance level. As Berger and Ofek (1995) document, diversified firms tend to trade at significant value discounts compared to their

Table 4. OLS Regression Analysis of Leverage Using Pooled Time-Series Cross-Section Data for Electric and Gas Utilities, 1980–1992.

See Table 1 for definition of variables and sources of data. T-statistics in parentheses. *, **, and *** denote statistical significance at 10%, 5%, and 1% levels , respectively.

Panel A: Depreciation / Total Assets (DPR/TA) used as a proxy for non-debt tax shield

Dep. Variable	Const.	MVE/BVE	INV/TA	AHERF	IGP/TA	OI/TA	LSALES	DPR/TA	NDT/TA	ITC/TA	SIGMA	Nobs	F-Stat (p-val.)	Adj. R^2 %
TD/TA	0.4302 (24.47)***	–0.0079 (–1.77)*	0.1950 (1.71)*	0.0727 (7.63)***	–0.0451 (–2.96)***	–0.5846 (–7.44)***	–0.0076 (6.06)***	–1.1240 (–3.84)***			0.0748 (3.36)***	1230	65.59 (0.00)	29.60
LTD/TA	0.2864 (17.14)	–0.0088 (–2.07)**	0.0608 (0.56)	0.0826 (9.12)***	0.0060 (0.42)	–0.2049 (–2.74)***	0.0112 (9.45)***	–1.9019 (–6.83)***			0.0575 (2.72)***	1230	63.75 (0.00)	29.00
TD/MVA	0.8515 (45.38)***	–0.1520 (–30.00)***	0.6844 (5.67)***	0.0298 (2.86)***	–0.1333 (–8.29)***	–1.1011 (–13.21)***	0.0004 (0.26)	1.3148 (4.20)***			0.0567 (2.45)***	1163	287.47 (0.00)	66.36
LTD/MVA	0.6148 (33.51)***	–0.1449 (–29.25)***	0.4804 (4.07)***	0.0602 (5.93)***	–0.0537 (–3.42)***	–0.5677 (–6.97)***	0.0071 (5.26)***	0.0636 (0.21)			0.0392 (1.74)*	1163	233.85 (0.00)	61.58
Predicted Sign		–	+	–	+	–	+	–			–			

Table 4. Continued.

Panel B: Imputed values (NDT/TA) used as a proxy for non-debt tax shield

Dep. Variable	Const.	MVE/BVE	INV/TA	AHERF	IGP/TA	OI/TA	LSALES	DPR/TA	NDT/TA	ITC/TA	SIGMA	Nobs	F-Stat (p-val.)	Adj. R^2 %
TD/TA	0.4307 (25.95)***	–0.0168 (–3.96)***	0.2814 (2.60)**	0.0810 (9.53)***	–0.4020 (–3.38)***	–0.7763 (–9.86)***	0.0055 (4.61)***		–0.0438 (–0.62)		0.0861 (4.24)***	1096	67.05 (0.00)	32.55
LTD/TA	0.3001 (18.15)***	–0.0192 (4.53)***	0.1223 (1.14)	0.1223 (11.94)***	–0.5552 (–1.13	0.1010)(–7.08)***	0.0087 (7.34)***		0.0437 (0.62)		0.0718 (3.55)***	1096	58.93 (0.00)	29.74
TD/MVA	0.8299 (41.31)***	–0.1382 (–25.59)***	0.7426 (5.73)***	0.0113 (1.08)	–0.1047 (–6.93)***	–0.9107 (9.74)***	0.0012 (0.80)		–0.0610 (–0.72)		0.0638 (2.69)***	1035	225.97 (0.00)	63.51
LTD/MVA	0.6091 (30.69)***	–0.1367 (–25.61)***	0.5177 (4.05)***	0.0593 (5.77)***	–0.0512 (–3.42)***	–0.6305 (–6.83)***	0.0067 (4.63)***		0.0435 (0.52)		0.0524 (2.23)**	1035	187.10 (0.00)	59.01
Predicted Sign		–	+	–	+	–	+		–		–			

Table 4. Continued.

Panel C: Investment Tax Credit / Total Assets (ITC/TA) used as a proxy for non-debt tax shield

Dep. Variable	Const.	MVE/BVE	INV/TA	AHERF	IGP/TA	OI/TA	LSALES	DPR/TA	NDT/TA	ITC/TA	SIGMA	Nobs (p-val.)	F-Stat %	Adj. R^2
TD/TA	0.4295 (23.35)***	–0.0151 (–3.32)***	0.4304 (3.81)***	0.1100 (11.00)***	–0.0231 (–1.75)*	–0.9198 (–12.58)***	0.0011 (0.86)			–5.1874 (–3.39)***	0.0676 (3.09)***	889	73.42 (0.00)	39.48
LTD/TA	0.2802 (15.29)***	–0.0205 (–4.52)***	0.2547 (2.26)**	0.1270 (12.68)***	0.0114 (0.86)	–0.6321 (–8.63)***	0.0069 (5.25)***			–5.4108 (–3.55)***	0.0554 (2.55)**	889	61.61 (0.00)	35.32
TD/MVA	0.8193 (35.67)***	–0.1351 (–23.62)***	0.9082 (6.58)***	0.0502 (3.89)***	–0.1071 (–6.60)***	–0.9437 (–10.65)***	–0.0023 (–1.37)			–3.1841 (–1.74)*	0.0377 (1.45)	842	197.21 (0.00)	65.11
LTD/MVA	0.5644 (24.83)***	–0.1370 (–24.21)***	0.6989 (5.12)***	0.0997 (7.80)***	–0.0389 (–2.42)**	–0.6037 (–6.89)***	0.0064 (3.90)***			–4.0959 (–2.26)**	0.0256 (0.99)	842	161.47 (0.00)	60.42
Predicted Sign		–	+	–	+	–	+		–	–	–			

more focused counterparts. One possible explanation of their results is that diversified firms tend to be mismanaged because of high degree of informational asymmetry (between insiders and outsiders, as well as among divisions), and they tend to be associated with significant agency costs (possibly due to greater problems of observability and also divisional rivalries). The results regarding the relationship between diversification and leverage in this study seem to support findings of Berger of Ofek (1995). The positive coefficient of AHERF indicates that better managed firms are indeed the more focused firms, and that they are in a more favorable position to obtain debt financing.

The most unexpected finding in Table 4 concerns SIGMA, the measure of variability. In contrast to the expected negative sign, we have a significant positive coefficient for SIGMA in many regression models. We do not obtain the predicted negative sign in any of the specifications.

3.5. Risk

A major discrepancy from theory appears to arise in the case of risk. We find unexpected regression coefficients for both of our proxies for risk - AHERF and SIGMA. (Note that multicollinearity is not an issue here, since both of the coefficients are significant).

These findings highlight certain aspect of risk as a determinant of leverage. The theoretical literature has focused on the direct effect of risk, especially through bankruptcy costs, on leverage. However, risk has also an indirect effect that may suggest a positive relation with leverage. In particular, other things equal, risk is likely to reduce the value of non-debt tax shields. Majority of those tax shields (depreciation, tax credits) arise as a consequence of the nature of company's business and company's performance, rather than as a matter of deliberate choice. As a result, non-debt tax shields can be treated as exogenous in the context of optimal capital choice. If risk reduces the value of non-debt tax shields, the company has a greater need for more leverage (to secure alternative, interest-related, tax shields). Risk and leverage thus indeed can be positively related, and our findings affirm this hypothesis.

4. CONCLUSION

An examination of the book and market leverage of one industry - electric and gas utilities - reveals substantial changes in the use of debt. In particular, over the period from 1975 to 1994, total debt to assets and long-term debt to assets fell by about 10% each. Following the theoretical literature, we attempt to explain this substantial change in terms of determinants of leverage that have already been proposed - bankruptcy costs, growth, variability, non-debt tax shields, collateral value, profitability, and size. While our empirical evidence is largely supportive of predicted effects of those determinants, the findings regarding the relation between leverage and variability (risk) are highly unexpected. We find that leverage and our risk proxies are positively related, whereas theoretical arguments based on the existence of bankruptcy costs and other considerations predict a negative relation. These conflicting findings point to potentially different effects that risk may have on the determination of leverage. The direct effect of risk (for example, through the increasing probability of bankruptcy) may indeed have a negative impact on optimal choice of debt. However, risk may have indirect effect as well. Specifically, risk likely reduces the value of non-debt tax shields. Lower non-debt tax shields, in turn, should be associated with a need to increase leverage (in order to secure interest-related tax shields). This effect of risk on debt has not been recognized in the previous literature, but it is consistent with the empirical findings of this paper.

NOTE

1. It is assumed that non-debt tax shields (investment tax credits, depreciation, etc.) are exogenous to the extent that they arise predominantly as a consequence of the nature of a firm's business. Thus, for a given amount of non-debt tax shields, a firm with greater risk runs a greater chance of finding that its non-debt tax shields are at an inadequate level. So, as compensation, the firm reacts with greater leverage.

ACKNOWLEDGMENTS

We are grateful to Howard Thompson for helpful discussions that have substantially improved this paper.

REFERENCES

Altman, E. I. (1984). A further empirical investigation of bankruptcy cost question. *Journal of Finance*, *39*, 1067–1089.

Ang, J. A., Chua, J. H., & McConnell, J. J. (1982). The administrative costs of corporate bankruptcy: A note. *Journal of Finance*, *37*, 219–226.

Berger, P. G., & Ofek, E. (1995). Diversification's effect on firm value. *Journal of Financial Economics*, *37*, 39–66.

Boquist, J. A., & Moore, W. T. (1984). Inter-industry leverage differences and the DeAngelo-Masulis Tax Shield Hypothesis. *Financial Management*, *13*, 5–9.

Bowen, R. M., Daley, L. E., & Huber, C. (1982). Leverage measures and industrial classification: Review and additional evidence. *Financial Management*, *11*, 10–20.

Bradley, M., Jarrell, G., & Kim, E. H. (1984). On the existence of an optimal capital structure: Theory and evidence. *Journal of Finance*, *39*, 857–878.

Chaplinsky, S., & Niehaus, G. (1993). Do insider ownership and leverage share the same determinants? *Quarterly Journal of Business and Economics*, *32*, 51–65.

DeAngelo, H., and Masulis, R. W. (1980). Optimal capital structure under corporate and personal taxation. *Journal of Financial Economics*, *8*, 3–29.

Harris, M., & Raviv, A. (1990). Capital structure and the informational role of debt. *Journal of Finance*, *45*, 321–349.

Harris, M., & Raviv, A. (1991). The theory of capital structure. *Journal of Finance*, *46*, 297–355.

Jensen, M. (1986). Agency costs of free cash flow, corporate finance, and takeovers. *American Economic Review*, *76*, 323–339.

Jensen, M., & Meckling, W. (1976). Theory of the firm: Managerial behavior, agency costs, and ownership structure. *Journal of Financial Economics*, *3*, 305–360.

Kester, C. W. (1986). Capital and ownership structure: A comparison of United States and Japanese manufacturing corporations. *Financial Management*, *15*, 5–16.

Kim, W. S., & Sorensen, E. H. (1986). Evidence on the impact of the agency costs of debt in corporate debt policy. *Journal of Financial and Quantitative Analysis*, *21*, 131–144.

Modigliani, F. & Miller, M. H. (1958). The cost of capital, corporation finance, and the theory of investment. *American Economic Review*, *48*, 261–297.

Modigliani, F., & Miller, M. H. (1963). Taxes and cost of capital: A correction. *American Economic Review*, *53*, 433–443.

Myers, S., & Majluf, N. (1984). Corporate financing and investment decisions, when firms have information investors do not have. *Journal of Financial Economics*, *13*, 187–221.

Navarro, P. (1982). Public utility commission regulation: performance, determinants, and energy policy impacts. *The Energy Journal*, *3*, 119–139.

Rajan, R. G., & Zingales, L. (1995). What do we know about capital structure? Some evidence from international data. *Journal of Finance*, *50*, 1421–1460.

Scott, J. (1977). Bankruptcy, secured debt, and optimal capital structure. *Journal of Finance*, *32*, 1–20.

Smith, C. W., & Watts, R. L. (1992). The investment opportunity set, and corporate financing, dividend, and compensation policies. *Journal of Financial Economics*, *32*, 263–292.

Titman, S., & Wessels, R. (1988). The determinants of capital structure choice. *Journal of Finance*, *43*, 1–20.

Warner, J. B. (1977). Bankruptcy costs: Some evidence. *Journal of Finance*, *32*, 337–347.

Weiss, L. A. (1990). Bankruptcy resolution: Direct costs and violation of priority claims. *Journal of Financial Economics*, *27*, 285–314.

EFFECTS OF HARMFUL ENVIRONMENTAL EVENTS ON REPUTATIONS OF FIRMS

Kari Jones and Paul H. Rubin

ABSTRACT

Many previous event studies have found unexpectedly large losses to firms involved in negative incidents. Many of these studies' authors explain such losses as "goodwill losses" or "reputation effects." To test this hypothesis, we search for residual losses (in excess of direct costs) to firms involved in events that produce ill will, but do not affect the quality of the firms' final products nor break implicit labor or supply contracts. Our sample of events is 73 negative environmental events reported in the Wall Street Journal *between 1970 and 1992 in which electric power companies or oil firms with listed stocks were involved. We find an overall insignificant capital market response. We interpret this as showing that firms are punished only for actions that actually harm customers or suppliers. Although others have found similar outcomes, our results enhance previous research by extending the findings to a broader range of environmental incidents over a longer time period. Further, our findings suggest that the large residual losses in other event studies may be due to reputation mechanisms (and not measurement errors or event study idiosyncrasies), when defined traditionally – only those who are (potentially) harmed incur the costs of punishment.*

Advances in Financial Economics, Volume 6, pages 161–182.
Copyright © 2001 by Elsevier Science B.V.
All rights of reproduction in any form reserved.
ISBN: 0-7623-0713-7

INTRODUCTION

In past event studies, researchers have found large unexplained capital market losses to firms involved in "negative" incidents. Implicitly and explicitly many authors attribute these residual losses to attrition of reputational capital, often for lack of another explanation. Theoretical models of reputation describe retaliation or punishment by a firm's contractual partners when the firm deviates from: an implicit agreement on quality of its product (punishment by consumers); an implicit labor contract (punishment by employees); an implicit purchasing agreement (punishment by suppliers); or a profit-maximizing strategy (punishment by shareholders). However, some of the authors of these studies, and those commenting on them, also imply that reputation may include punishment by the firm's contractual partners for harm done to others. Mainstream theoretical models of reputational mechanisms do not predict that one group will punish when a different group is harmed. Unless the group doing the punishing expects to be harmed if the firm's devious behavior continues, punishment requires that the harmed group's well-being enter the punishing group's utility functions.

The factors driving the unexpected results of previous event studies are yet to be explained. Policy decisions and academic questions depend on correctly explaining when and how reputational mechanisms affect the capital market. This paper provides evidence that reputational mechanisms are a plausible explanation of the unexplained residual losses found in previous papers, when reputation is defined traditionally. That is, only those who are (potentially) harmed incur the costs of punishment.

The remainder of this paper is organized as follows: the first section contains an overview of previous event studies germane to the question of reputation effects. The next section includes an explanation of these past results within the traditional models of reputation. Additionally, in this section, we show the prevalence in the academic literature and popular press of the assumption that a firm's social reputation affects its capital market value. We also outline an empirical test of this assertion. Next, we briefly discuss our empirical procedure and the data. Finally, we present the empirical results and conclusions.

UNEXPLAINED PAST EMPIRICAL FINDINGS

Unexplained losses have been found in a variety of previous event studies that were designed to measure the effect (and effectiveness) of regulation on a firm or industry.[1] More recently, similar effects have been found for an even wider variety of regulatory and non-regulatory events. Among the regulatory event studies, significant capital market losses are associated with firms' involvement in Federal Trade Commission censures for false and deceptive advertising (Peltzman, 1981; Mathios & Plummer, 1989), government-ordered drug, automobile, and other product recalls (Jarrell & Peltzman, 1985; Hoffer et al., 1988; Rubin et al., 1988; Bosch & Lee, 1994), other product-safety-related regulatory (and private) actions (Viscusi & Hersch, 1990), Equal Employment Opportunity (EEO) violations (Hersch, 1991), Occupational Safety and Health Administration (OSHA) violations (Davidson et al., 1989; Fry & Lee, 1989), and corporate crimes such as fraud and price fixing (Cloninger et al., 1988; Karpoff & Lott, 1993; Reichert et al., 1996).

Non-regulatory events also lead to large and often unexplained losses. These include changing of a product's formula (Benjamin and Mitchell, undated manuscript), the Tylenol poisonings (Mitchell, 1989; Dowdell et al., 1992), and airline crashes (Mitchell & Maloney, 1989; Chalk, 1987). While these studies show the negative effects of a bad reputation, one study provides evidence that *good* reputations can *increase* a firm's value. Chauvin and Guthrie (1994) found that firms experienced a statistically significant average gain in market value from appearing on *Working Mother* magazine's list of "best" employers.

In most of these studies the monetary losses to stockholders significantly outweigh the direct and estimated indirect costs of the incidents. These results are surprising to many authors. Peltzman (1981) characterizes his findings as "amazing" and ". . . a mystery" (at 418), while Rubin et al. (1988) label their extremely significant findings "surprisingly large" (at 37). For lack of a better explanation many authors characterize the residual losses (above and beyond explainable costs) as losses of reputation or goodwill. Dowdell et al. (1992), Jarrell and Peltzman (1985), and Rubin et al. (1988) all characterized the excess losses in their studies as losses of a firm's goodwill. Mitchell and Maloney (1989) dubbed their residual losses a "brand name effect."

Previous event studies involving environmental events show mixed results. Muoghalu, et al. (1990) found that in hazardous waste lawsuits that allege damages from improper hazardous waste disposal, defendant firms suffered significant losses. However, Harper and Adams (1996) found that the average market loss suffered by firms named as a potentially liable party in a Superfund cleanup effort was not significantly different from zero. Laplante and Lanoie (1994) found that for negative environmental events reported in the media, Canadian-owned firms did not experience significant declines in stock market value either when an environmental violation was announced or when a lawsuit was filed. This is consistent with the small average penalty paid. The authors found that significant market adjustment occurred only after a suit settlement was announced. It is not known if the firms experienced residual losses.

Karpoff, Lott and Rankine (1998) found a statistically significant average loss of 0.85% to firms involved in negative environmental incidents. The average loss was greater for events where initial press reports of the incident occurred either at the allegation date or the date charges were filed. However, when comparing these figures with the direct costs of the incidents, the authors found no evidence that any part of this loss could be attributed to reputation effects.

Hamilton (1995) investigated the stock market reaction to information releases concerning a firm's pollution activities. Manufacturing facilities must report annual releases of chemicals to the EPA. This information is relayed to the public in a database called the Toxics Releases Inventory (TRI). Hamilton studied both how the media treated TRI information and what affects it had on stock prices of polluting firms. He found that on the day of the information release, firms suffered, on average, a statistically significant drop in capital market value. This loss was greater the larger the number of chemicals per facility. Capital market losses were also positively correlated with number of Superfund sites with which the firm was involved. Hamilton's explanatory regressions suggested that potential liability and other direct monetary exposure issues, rather than consumer forces, drove these losses.

Finally, Blacconiere and Patten (1994) reported that Union Carbide took a 27.9%, or approximately $1 billion, hit from the Bhopal chemical leak, while its industry rivals suffered an average 1.28% loss in capital market value. The authors attributed these losses to investors' revisions of possible production-side risks and increased regulatory exposure. A summary of these event studies is presented in Table 1.

EXPLAINING PAST EMPIRICAL RESULTS

Traditional Theories of Reputation

The authors of some of the studies reviewed above attempt to explain their estimated capital market losses by regressing them on study-specific potential explanatory factors. Others offer ad hoc explanations of possible factors affecting the magnitude of losses. However, none attempts to formally model the reputational mechanism at work when the authors refer to "reputation effects".

Traditional theories of reputational mechanisms have their roots in the concepts that are articulated in Akerlof (1970), Klein, Crawford and Alchian (1978), Klein and Leffler (1981), Nelson (1970), and Nelson (1974), and are modeled formally in Shapiro (1983). Akerlof (1970) noted that in some

Table 1. Summary of Previous Event Studies.

Event Type	Average Loss in $	Average Loss as Percentage of Value	Cite(s)
FTC cease-and-desist order for false advertising	as high as $250M for larger firms	3.2–6.4	Peltzman, 1981; Mathios & Plummer, 1989
Automobile recalls	n.a.	insignificant	Jarrell & Peltzman, 1985; Hoffer et al., 1988
FDA-mandated drug recalls	$1.8M (residual)	5.6 (residual)	Jarrell & Peltzman, 1985
FDA disciplinary actions	n.a.	2.22–3.42	Bosch & Lee, 1994
Drug packaging regulations (subsequent to Tylenol poisonings)	$310M	11.83	Dowdell et al., 1992
CPSC-mandated recalls	$146M	5.4–6.9	Rubin et al., 1988
EEO violations; EEO class-action suits	$18.5M $16M (residual)	0.29–0.48; 15.6	Hersch, 1991
OSHA sanctions	$3.9–22.1M	0.53–2.1	Fry & Lee, 1989; Davidson et al., 1994
Corporate indictments	$9.15M (residual; not sig.)	1.38–3.1	Reichert et al., 1996
Charges and pleas in price-fixing cases	n.a.	17	Cloninger et al., 1988
Hazardous waste mismanagement suits	$33.3M	1.2	Muoghalu et al., 1990
Environmental violations	n.a.	0.85 net loss to full sample; no residual	Karpoff et al., 1998
Being named a potentially responsible party at a Superfund site	n.a.	insignificant	Harper & Adams, 1996

Table 1. Continued.

Event Type	Average Loss in $	Average Loss as Percentage of Value	Cite(s)
Violation of Canadian environmental regulation	n.a.	insig. for announcement; 2.0 for settlement	Laplante & Lanoie, 1994
Coca-Cola adopts New Coke	$500M	8.1	Benjamin & Mitchell
Criminal fraud against stakeholders and government	$40–60.8M	1.34–5.05	Karpoff & Lott, 1993
Tylenol poisonings (first wave)	$1.24B (residual)	14.3 (residual)	Mitchell, 1989
Bhopal, India chemical leak	~$1B (Union Carbide) n.a. (Union Carbide's rivals)	27.9 (Union Carbide) 1.28 (Union Carbide's rivals)	Blacconiere & Patten, 1994
Chemical releases reported	$4.1M	0.28	Hamilton, 1995
Airline crashes before and after deregulation	$11–18.3M (residual)	1.34–1.55 (residual)	Mitchell & Maloney, 1989
Aircraft-manufacturer-at-fault crashes	$21.32M	3.8	Chalk, 1987
Included in *Working Mother*'s "best" employers list	n.a.	0.28–2.2	Chauvin & Guthrie, 1994

situations of asymmetric information between buyers and sellers, mutually beneficial trades may be precluded by the prospects of cheating. Subsequently, economists began describing and modeling the methods that have evolved in such markets to mitigate[2] this problem. In particular, firms often use reputation to guarantee product quality.

Klein et al. (1978) first suggested that potential cheaters might offer a forfeitable hostage to guarantee performance in interfirm contracting. Klein and

Leffler (1981) applied this concept to the consumer-producer relationship. High prices signal high quality, but consumers will be willing to pay these prices only if they receive some guarantee of high quality. Producers of high quality goods make firm-specific investments that are forfeited if consumers discontinue purchases of the firm's output. Consumers realize that firms are unlikely to deviate from high quality production, because by doing so they forfeit their investments in "hostages."[3]

Thus, if a firm deviates from a quality commitment (or even if an incident occurs that signals the probability that the firm has deviated), customers punish the firm by lowering their willingness to pay. The net present value of the future profit stream declines and the capital market value of the firm falls.[4] The firm also has a reputation in its dealings with suppliers and employees. If a firm cheats in some way on a commitment it has to a supplier, the firm is likely to incur higher input costs in the future. Firms that cheat their employees may face wage premiums to attract future workers in a competitive labor market. In each case, the value of the firm's future profits decreases.

Finally, firms also have reputations with investors. A firm's involvement in a negative incident may lead investors to reevaluate their faith that the firm's management will steer clear of such costly events in the future. Investors also may revise upward their subjective probabilities of tighter future regulatory scrutiny or additional regulations. The higher the perceived risk of such consequences, the lower the firm's expected future profit stream, and the lower the share price.

While these traditional interpretations of reputational factors are reasonable partial explanations of observed residual losses, they leave two issues unresolved. First, no empirical evidence exists to show that the combination of direct costs and these reputational penalties explains all of the losses reported in empirical studies. In fact, both Peltzman's work (1981), and Peltzman's and others' comments suggest otherwise. Second, suggestions of the importance of social reputation in explaining these residual losses are pervasive in the literature. That is, it is widely suggested that the firms' contractual partners punish firms for harm done to others. Without a unified model of reputation (consistent with the stylized facts), the social reputation hypotheses, while not compatible with economic theory, continues to carry as much weight as the other ad hoc explanations of residual losses.

The (Conjectured) Social Component of Reputation

The foundations of the case for social reputation are found in suggestions and anecdotes in the literature and also in the results of experimental economics.

Many of the original event studies' authors merely suggest that a firm's social reputation can affect sales.[5] Some quotes are more direct. Hersch (1991) states that in the aftermath of an EEO violation "costs include . . . adverse publicity that might result in the loss of sales . . ." (at 140). Muoghalu et al. (1990), in reference to residual losses suffered by firms involved in Superfund cases, state that "stockholder losses . . . include . . . public ill-will resulting from the lawsuit or the dumping" (at 358, note 5). Hanka (1992) notes "image-conscious firms fear the reputation consequences of pollution" (at 26). Davidson et al. (1994) attribute some of the significant losses suffered by violators of OSHA regulations to "negative publicity for the firm".

These comments suggest that consumers, employees, or suppliers[6] punish firms for engaging in practices that are "socially irresponsible." The losses suffered from this type of retaliation would be in addition to other direct and indirect costs of the incident, including reputational losses for failing to honor implicit contracts with consumers, employees, or suppliers, and any losses from decreased faith in management. Thus, such "social" punishments would create the unexplained "goodwill" or "reputation" effects.

Suggestions of market effects from social reputations are not limited to punishment for bad reputations, but extend to rewards for good reputations. Chauvin and Guthrie (1994) state ". . . if investors believe that customers will prefer purchasing goods and services from 'good' employers, [the positive returns to firms on *Working Mother*'s 'best employer' list] may also reflect estimates of the effect that labor market reputation may have on sales" (at 551). Schwartz (1968) states that "(g)ifts which enhance the public image of a corporation can advantageously shift the demand curve for the corporation's product" (at 480). Navarro (1988) concluded that corporate donations to charity increase demand, or decreases demand elasticity, for a firm's product.

Anecdotal evidence that customers gain utility and disutility from characteristics of firms' production and financing processes feeds the "social reputation" hypothesis. Consumers readily purchase recycled products (such as notebook paper and paper towels), which function no better than nonrecycled products, yet command price premiums. Investors put money in socially conscious mutual funds that pay lower returns for the risk than their socially-disinterested counterparts. Rothchild (1996) states that "[o]ver the past 12 months, the 39 ethical funds tracked by Lipper [Analytical] have returned 18.2% vs. 27.2% for the S&P 500" (at 197). Rogers (1996) translates this into a $57.5 billion price investors were willing to pay to avoid investing in undesirable firms over the year.

Observation of business practices like voluntary divestiture from South Africa, advertising to tout environmental friendliness or "fairness" in hiring practices, appointment of environmentalists to boards of directors, and emphasis in stockholder reports on politically correct firm policies suggest that firms reap some reward from consumer or investor knowledge of these practices. Such practices do not improve the efficiency of the actual production process, but, because they are common practice, firms must expect that they add to profits. A spokesman for Reebok International Ltd. claims "[m]ore and more in the marketplace, . . . who you are and what you stand for is as important as the quality of the product you sell" (Hayes & Pereira, 1990, at B1, B7).

Adding to the fervor of these suggestions of social reputation are experimental economists' reports of evidence of widespread principle-based behavior in laboratory tests[7] and behavioral models in the literature that are generated to be compatible with these experimental results.[8]

TESTING FOR EVIDENCE OF A SOCIAL REPUTATION EFFECT

Both the formal definitions of reputation in the literature (see, for example, Shapiro, 1982) and the formal discussions in the above event studies assert that "reputation effects" refer only to reputational mechanisms that involve self-interested punishment.[9] However, the informal discussion suggests that social reputation affects profit. These assertions have significant consequences. Voters, regulators, and taxpayers make assumptions concerning retaliatory behavior trends when voting for (or otherwise affecting) government involvement in markets. For example, Karpoff et al. (1998) note that the U.S. Sentencing Commission is explicitly considering the existence of reputational effects from environmental incidents in setting its sentencing guidelines for corporate environmental crimes.

The goal of the empirical section of this paper is to test whether, as copious comments suggest, social reputations are a factor in the unexplained losses of previous event studies. Our procedure is to test for residual losses (net of direct costs) to firms involved in negative events that do not harm the firm's contractual partners (except through direct losses from the incident), but do affect their social reputation by harming third-parties. Negative environmental incidents represent such events. Absence of residual losses suggests that agents do not punish firms for harm done to others, while presence of significant residual losses suggests that measurable reputation effects may result from negative social reputation.

Before undertaking this empirical investigation, we expected to find significant residual losses, based on the results of similar event studies. We noted that if we found residual losses, we would need to show that these losses represent punishment for bad social reputation, and not, for example, a measurement problem common to all event studies. As such, we structured our empirical test so that this question could be answered. We conjectured that consumers' propensities to punish for social reasons decrease as the cost of punishment increases, but a measurement error or other extraneous factor should not vary with costs of punishment. We collected a sample of environmental mishaps caused by electric power companies and oil companies. The utilities represent an industry with few substitutes (where punishment would be expensive and, therefore, less likely), while the oil companies represent an industry with many close substitutes (where punishment would be relatively cheaper and, therefore, more likely). Thus, we could compare residual losses between cheap-to-punish and expensive-to-punish industries if needed.

THE EMPIRICAL MODEL AND PROCEDURE

The Event Study Data and Methodology

A list of potential events was drawn from all negative environmental events involving oil concerns and electric utilities between 1970[10] and 1992[11] as reported in *The Wall Street Journal Index*. Events with other firm-specific news during the event window were removed. Potential events had to meet two additional criteria. First, the event must have had a negative environmental impact as the result of the actions of an oil division or electric power producing division of the firm. This is because, if consumers were distressed upon hearing of pollution by Acme Chemical, but were unaware that Acme was a subsidiary of ABC Oil Co., they would be unable to punish, if so inclined. Second, the event must not have affected the quality of the firm's physical product. For example, some oil firms have been charged with switching leaded and unleaded fuel. Using the wrong type of fuel not only causes pollution, but also inflicts costly damage to a car's engine and exhaust system. In such cases, consumers' material self-interest would lead to decreased demand; retaliation for any resulting pollution could not be separated.

We use as our event date, the day that news of the event first appears in the *Wall Street Journal* (*WSJ*), unless the report suggests a more appropriate date. In picking our event dates, we carefully considered the fact that if *WSJ* announcement dates are used as event dates when they do not correspond to the date of the precipitating incident, the market may have adjusted to news of the incident before the event date, consequently biasing announcement date losses downward.

The information to which market participants are most likely to react is not always the precipitating incident. The market reaction to an announcement of an investigation than has, say, a 5% chance of resulting in a $1M fine will be much different that the reaction to the commencement of an investigation that has, say, a 90% probability of resulting in a $1M fine. The best option is to use the date most likely to contain the bulk of the market adjustment to the "event."

We feel that the *WSJ* announcement date is the appropriate event date for our study for three reasons. First, most of the other studies with which we compare our results employ this methodology. To allow meaningful comparisons of our results with past studies' results, we follow comparable procedures. Second, there is precedent for using this methodology. Many event studies have used the *WSJ* announcement date as the event date, even when it did not necessarily correspond to the date of the precipitating incident, and still found significant market reactions to the announcement news. This suggests that the *WSJ* mainly reports news that is likely to signal a significant probability of a change in a firm's value.

Thus, the third reason we use the *WSJ* report date as our event date is that we feel the *WSJ* gets it right most of the time – its reports of news that affects the market are reasonably accurate and timely. At a minimum, there is enough uncertainty left when the *WSJ* publishes a news story that the information is still novel. The results of previous event studies (particularly previous environmental event studies), and the seemingly-random actions of the EPA and environmental groups, suggest that being implicated in an environmental incident often is associated with a low probability of further action or concern. Karpoff, Lott and Rankine (1998), at 25, conclude that penalties stemming from environmental incidents are highly variable and not easy to predict. Thus, news of many potential events is unlikely to be printed until some official action is taken. However, some events which seem highly likely to generate further costs (e.g. Valdez and Three-Mile Island) were reported very close to the precipitating incident date.[12]

Initially, we found 98 events meeting our three criteria, but removed 25 events with concurrent news in the event window, producing a final sample of 73 events. The average capital market value of the firms at the time of the incident was approximately $7.9 billion, with a low of $38.4 million and a high of $72.6 billion. 53 firms are oil companies and 20 are electric utilities.

We employ standard event study methodology to identify firms' abnormal returns from negative environmental incidents. The abnormal return accruing to a firm as the result of an event is estimated as the difference between the firm's actual return at the time of the event and the return that would be expected for the firm in the absence of the incident. For any given day, the abnormal return is calculated:

$$AR_{it} = RET_{it} - (a_i + b_i MRET_t)$$

where AR_{it} is the abnormal return to the target firm of event i on day t, RET_{it} ≡ the percentage return to the target firm of event i at date t, and $MRET_t$ ≡ the percentage return at time t to the NYSE/AMEX value-weighted portfolio. Time, t, is indexed such that t = 0 on the event date.

We employ the market model to obtain parameters for generating expected returns. Specifically, for each event, using returns for 199 days just prior to the event date, we obtain a_i and b_i as estimates of α_i and β_i from an OLS estimate of:

$$RET_{it} = \alpha_i + \beta_i MRET_t + e_{it}$$

where e_{it} is random disturbance to the event i target firm's return at date t. For each event, the e_t are assumed to be distributed with a mean of zero and variance of σ_i^2.

The average abnormal return to the entire sample on any day t, denoted AAR_t, is the average of the abnormal returns on day t for each event in the sample:

$$AAR_t = \frac{\sum_{i=1}^{n} AR_{it}}{n},$$

where n is the number of events in the sample.

The abnormal return to the firm accumulated over the period that the market adjusts to the news is our best estimate of the losses (or gains) to the firm from the event.[13] For a given event window [e.g. the event window consisting of days t = −1 and t = 0, denoted (−1,0)], the cumulative abnormal return for event i, denoted CAR_i, is defined as the sum of the abnormal returns for each day in the event window:

$$CAR_t = \sum_{t=f}^{l} AR_{it}$$

where f ≡ the first day in event i's event window, and l ≡ the last day in event i's event window.

For any given event window [e.g. the (−1,0) window], the average cumulative abnormal return to the full sample, ACAR, is the average over all events of each CAR_i:

$$ACAR = \frac{\sum_{i=1}^{n} CAR_i}{n}.$$

For a given event window, the total monetary loss to the firm's shareholders from the event is simply the product of the firm's CAR and the value of the firm's shares outstanding on the day $t = -2$.

Cross-Sectional Analysis of the Event Study Results

After we calculate the cumulative abnormal returns from each event, we investigate the factors contributing to these observed abnormal returns. The (−1,0) window CARs for all events are regressed on event-specific explanatory variables in an attempt to identify other causal factors that may explain the event study results. The first explanatory variable is report type. This tests whether timing of the news report systematically affects observed returns. The second explanatory variable, action type, accounts for the possibility that different types of accusers impose different costs on polluters.

The third explanatory variable is severity of the event. This tests whether an overwhelming number of trivial events are watering down the reaction to severe and significant events when CARs are averaged over the entire sample. The fourth explanatory variable is time. It may proxy for environmental consciousness and regulatory and/or prosecutorial fervor in the environmental sector. In an alternative specification we substitute presidential administration dummy variables for time.

The fifth explanatory variable is industry. Dummying by industry controls for the possibility that oil firms' and electric firms' reactions to new information are systematically different, either because the utilities are rate-of-return regulated, and/or because their stocks are traded more thinly, on average, than oil stocks. The sixth and final explanatory variable, firm size, controls for the possibility that extremely large firms' abnormal returns test insignificant because losses from the incident are dwarfed by the sheer magnitude of the firms' market values.

The explanatory regression results are obtained by an OLS estimate of

$$\begin{aligned} CAR_i = {} & \alpha + \beta_1 REPORTC_i + \beta_2 REPORTR_i + \delta_1 ACTIONF_i + \delta_2 ACTIONP_i \\ & + \delta_3 ACTIONM_i + \tau_1 SMALL_i + \tau_2 MEDIUM_i + \tau_3 COMPL_i + \eta TIME_i \\ & + \varphi IND_i + \gamma SIZE_i + e_i \end{aligned}$$

where

REPORTC_i = 1 if first news of event i is reported at the time of the filing of the first suit, the first official allegation, the first official warning of an impending suit or charge, or a settlement concurrent with official charges,
= 0 otherwise.

REPORTR_i = 1 if first news of event i concerns a present court case, a judge or jury ruling, a penalty ruling, or a settlement (not concurrent with the accusation),
= 0 otherwise.

ACTIONF_i = 1 if event i involves only a federal action,
= 0 otherwise.

ACTIONP_i = 1 if event i involves only a private action,
= 0 otherwise.

ACTIONM_i = 1 if event i involves a multi-party action,
= 0 otherwise.

SMALL_i = 1 if the reported estimated or actual costs of the incident were less than $1M,
= 0 otherwise.

MEDIUM_i = 1 if the reported estimated or actual costs of the incident were >$1M, but <$100M,
= 0 otherwise.

COMPL_i = 1 if the reported direct costs were small, but the firm faced uncertain compliance costs,
= 0 otherwise.

TIME_i = the last two digits of the year of event i,

IND_i = 1 if event i involves an oil firm,
= 0 if event i involves an electric utility, and

SIZE_i = the total dollar value of firm i's outstanding shares at $t = -2$, measured in thousands of 1987 dollars.

RESULTS AND CONCLUSIONS

Abnormal and Cumulative Abnormal Returns[14]

A summary of the average abnormal returns (AARs) and the average cumulative abnormal returns (ACARs) for various windows is presented in Table 2. The split of negative and positive returns is not significantly different from what may occur randomly, and relatively few of the firms' stocks reacted significantly to news of the incident.

Table 2. Summary of Results for Various Windows.

All events:

	t = −1	t = 0	(−1,0)	(−5,0)[a]	(−1,9)[b]
Significantly-positive-to-significantly-negative ratio	3 : 5	4 : 5	4 : 7	5 : 4	4 : 7
Total-positive-to-total-negative-ratio (Probability of getting the actual split or even more skewed, assuming a 0.5 probability of a positive return)	33 : 40 (0.48)	42 : 31 (0.24)	38 : 35 (0.82)	35 : 38 (0.82)	35 : 38 (0.82)
AAR or ACAR (z-statistic)	−0.0031 (−2.0971)‡	0.0030 (1.5116)	−0.0001 (−0.4144)	0.0012 (0.0897)	−0.0013 (−0.9183)
Average change in $ value ($K)	−4341	16136	11800	−2924	40282

Without one extreme event and two events with industry-specific concurrent news:

	t = −1	t = 0	(−1,0)	(−5,0)a	(−1,9)[c]
Significantly-positive-to-significantly-negative ratio	3 : 4	1 : 5	2 : 6	4 : 3	3 : 7
Total-positive-to-total-negative-ratio (Probability of getting the actual split or even more skewed, assuming a 0.5 probability of a positive return)	32 : 38 (0.55)	39 : 31 (0.40)	36 : 34 (0.90)	33 : 37 (0.72)	32 : 38 (0.55)
AAR or ACAR (z-statistic)	−0.0012 (−0.9145)	−0.0004 (−0.2923)	−0.0015 (−0.8498)	0.0007 (0.0310)	−0.0044 (−1.3682)
Average change in $ value ($K)	−3855	10052	6203	−8038	35483

‡ Significant at 5% level.

[a] These are estimates, because the estimation period and event window overlap for these calculations.

[b] Because of data restrictions, a (−1,5) window had to be used for one event and (−1,8) used for two others.

[c] Because of data restrictions, a (−1,5) window had to be used for one event and (−1,8) used for another.

When all events are considered, the ACAR to the full sample for the (−1,0) window is an insignificant −0.0001. Additionally, there are four significantly positive CARs. These results are counterintuitive, because, even if reputation effects are not present, there are direct costs to the incidents. Two events shared an event date on which Mideast tensions threatened to delay a possible loosening of oil importing restrictions by President Nixon. Both events' CARs are significantly positive. Another event involved a radiation leak scare at a Rochester Gas and Electric plant only a few years after Three Mile Island. At first news of the incident (t =−1), a mass sell-off resulted in a 13.5% drop in the firm's value. However, when more information was released the following day (t=0), the stock gained back almost nine percent (and the stock's price was back to its pre-event level within five trading days). This event alone seems to be driving the significant ACAR at t =−1. Removal of all three of these events leaves an insignificant average market loss to the sample of −0.0015 for the (−1,0) window. In dollar terms, the firms actually gained $6.2M, on average, during the (−1,0) window. These figures represent gross losses (and gains). Residual losses were not calculated since the ACAR is insignificant. That is, netting out the losses due to direct costs can only make an individual event's return a larger (less negative) number. Thus the ACAR net of direct costs cannot be significantly less than zero.

Insignificant average cumulative abnormal returns in the (−5,0) window and (−1,9) window suggest that the stock market's adjustment to the events did not occur outside of the (−1,0) event window. The insignificant (−5,0) window suggests that news of the events did not reach the market just prior to the report date, while the insignificant (−1,9) window suggests that adjustment to the new news did not occur with a lag.

Explanatory Regression Results

Our event study results suggest that abnormal returns to firms involved in negative environmental events are random and average approximately zero. Compared to event studies of other negative events that employ a similar methodology, we find a much smaller effect from negative environmental incidents. To test whether some causal variable(s) could explain the variation of CARs over the different firms and incidents, we regress the observed cumulative abnormal returns for the (−1,0) window on explanatory variables for report type, action type, severity of the event, time, industry, and firm size.

A summary of the results of our explanatory regression is reported in Table 3. Consistent with our event study results, the explanatory regression also

Table 3. Explanatory Regression Results.

Intercept term or Coefficient on:	Estimated Value (t-statistic – 61 d.f.)	Estimated Value (t-statistic – 58 d.f.)
Intercept term	0.07294	0.01987
	(1.884)§	(1.003)
REPORTC	0.00514	0.00314
	(0.4918)	(0.2802)
REPORTR	0.00797	0.00508
	(0.6105)	(0.3809)
ACTIONF	−0.00158	−0.00217
	(−0.1927)	(−0.2594)
ACTIONP	−0.00881	−0.01127
	(−0.6241)	(−0.7629)
ACTIONM	−0.01380	−0.01804
	(−0.9135)	(−1.124)
SMALL	−0.00680	−0.01181
	(−0.5430)	(−0.8703)
MEDIUM	−0.01421	−0.01784
	(−1.251)	(−1.368)
COMPLIANCE	−0.01409	−0.01864
	(−1.100)	(−1.224)
TIME	−0.00091	–
	(−2.136)‡	–
FORD	–	0.00123
	–	(0.0863)
CARTER	–	−0.01782
	–	(−1.363)
REAGAN	–	−0.01714
	–	(−1.863)§
BUSH	–	−0.01383
	–	(−1.481)
IND	0.00561	0.00182
	(0.7272)	(0.2052)
SIZE	0.10281E−9	0.7477E−10
	(0.5689)	(0.4048)
F	1.248	1.071
R^2	0.0365	0.0136

§ = Significantly different from zero at 0.1 level of significance.
‡ = Significantly different from zero at 0.05 level of significance.

suggests that abnormal returns to firms are random and can be explained as white noise. The adjusted R^2's for the alternative specifications are less than 0.04 and only the variable for time has any significant explanatory power. The breakdown of the time variable into dummies for presidential administration shows that a good deal of this result can be explained by the unusually high percentage of costly events occurring during the Reagan administration. The Durbin-Watson statistics are 2.05 and 2.1 for the regressions incorporating year and presidential administration dummies, respectively, and a plot of the regression residuals appears normal.

The insignificance of the report type variables suggests that use of event dates postdating the precipitating incident date (which is often not possible to pinpoint) does not significantly bias the sample average cumulative abnormal return (ACAR) upward. This is accentuated by the fact that 59 events with first news at the time of the charge are being compared to only 14 events (including both the Valdez spill and Three Mile Island) with first news at the time of the precipitating incident. The insignificance of the coefficient on the dummy for industry suggests that oil and electric firms did not have significantly different market reactions to the environmental news due to regulatory or thin-trading issues.

If, as some authors have suggested, there were a reputational penalty to negative environmental events, we would expect to find significantly negative residual losses on average. Instead, the cumulative abnormal returns appear to be randomly distributed about zero. In keeping with the standard models of reputation, we find no evidence of a negative social reputation effect.

Our results are consistent with Harper and Adams's (1996) finding that firms experienced insignificant changes in value upon being named a potentially responsible party in a Superfund cleanup. They are also consistent with Laplante and Lanoie's (1994) finding that Canadian firms did not experience significant losses from announcements of environmental violations or lawsuits. Our results, however, extend this finding of insignificant[15] market losses in environmental event studies to a broader class of regulatory as well as nonregulatory events.

Our results are also consistent with the assertions and findings of Karpoff et al. (1998) that formal penalties for committing environmental crimes are random and that stockholders realize this. Because our sample covers a period twice as long as Karpoff et al.'s, (1970 through 1992 versus 1980 through 1991) we can be confident that their findings are not merely a figment of, for example, the "1980s mentality," but span several decades and presidential administrations.

CONCLUSIONS

We find an overall insignificant stock market response to a sample consisting of all (usable) negative environmental incidents (regulatory and nonregulatory) reported in the *Wall Street Journal* over the 1970 to 1992 period involving oil firms and electric power companies with listed stocks. Thus, our results suggest that other researchers' findings of insignificant reputation effects from regulatory environmental events extend to a wider class of regulatory and nonregulatory incidents than previously has been considered in the literature. These events affect firms' social reputations, but not the quality of their output or their reputations with employees or suppliers. Thus, our findings contradict the widely-asserted hypothesis that when firms develop negative social reputations (that is, negative reputations from harming third parties), their unaffected contractual partners will incur the costs of punishment. Rather, our findings affirm the traditional models of reputational mechanisms, which are effective when harms and punishments are limited to contractual partners. Moreover, our findings provide evidence that the large losses from harmful events are, in fact, due to stock market expectations that contractual partners will punish firms for imposing direct harms.

NOTES

1. See Schwert (1981) for an overview.
2. As Shapiro (1982) notes, full-information quality levels are unattainable even with a reputational mechanism.
3. Other investigations into the nature of the outcomes of various reputation mechanisms are found in the signaling models of game theorists. See, for example, Allen (1984), Kreps and Wilson (1982), Milgrom and Roberts (1986), and Kihlstrom and Riordan (1984).
4. Ippolito (1992) reported in a study of the mutual fund market that consumers rationally react in this manner, thus preventing a "lemons" market.
5. In contrast, Karpoff and Lott (1993) find that firms that are penalized for committing frauds that do not affect consumers, suppliers, or stockholders (such as paperwork errors) suffer no unexplainable stock market losses.
6. While socially conscious investors may purge the stock of firms they consider irresponsible from their portfolios, this phenomenon is unlikely to affect the observed stock market loss. In an efficient market, the investors who are interested only in risk and return will bid against each other for the divested shares until the price reflects the value of the firm.
7. See, for example, Camerer and Thaler (1995) and Kahneman, et al. (1986) in which players are willing to sacrifice money to reward players who are kind to others in previous rounds and punish those who were previously "unfair" to their opponents.

8. See, for example, Bolton (1991), in which relative payoffs matter, and Rabin (1993), in which agents may reward those who are kind and punish those who are unkind.

9. Chauvin and Guthrie (1994) note that "... in all theoretical work on reputations, reputations have economic value because they improve the efficiency of markets ..." (at 546). Schwartz (1968) notes that "(h)istorically, economists have tended to ignore private philanthropic behavior and to regard it as economically irrational" (at 479).

10. The environmental movement was gathering steam and the EPA was created this year.

11. Environmental fervor may be related to political climate and this year ended a presidential administration.

12. Additionally, to test for upward bias of abnormal returns resulting from late reports of important stories, we control for the timeliness of the report in our explanatory regression of these returns.

13. We assume that the market reacts immediately to news of the event. However, because it is unclear for many of our events whether this news reached the market before 4:00 p.m. on the day before it appeared in the *WSJ*, we use the (−1,0) event window. For comparison, we also calculate results for the day t=−1, day t=0, and the (−5,0) and (−1,9) windows.

14. A subset of fourteen events was used to test the effect of different combinations of estimation period (150-day or 200-day) and market return calculation (equally-weighted or value-weighted) specifications on the predictive power of the market model. No one measure or class of measures produced a significantly higher correlation between targets' returns and market returns. The subsample was also used to determine whether either a procedure to net out the effects of industry-wide increased regulatory scrutiny or a procedure to net out the residual losses due to decreased faith in management was warranted for the full sample. Neither was found to be of significant benefit.

15. Muoghalu et al. (1990) unsurprisingly found significant gross losses because they considered only a specific type of environmental incident that has (potentially large) direct costs. The authors did not report residual losses (net of direct costs), so reputation effects are not known. Hamilton (1995) found that firms suffered significant losses when information about chemical releases at their facilities was made public. However, in the author's explanatory regressions, losses increased with number of chemicals at a facility but not with level of emissions, indicating potential future liability was more important than extent of current harm. Also, number of Superfund sites was associated with increased loss, but extent of media coverage had an insignificant effect, again indicating that response to potential risk exposure outweighed reactions to the firms' social reputation and goodwill.

ACKNOWLEDGMENTS

We would like to thank Owen Beelders, Robert Carpenter, Joel Schrag, and Susan Griffin for providing helpful suggestions and comments. Additionally, we thank Xiaolan Wang of the Emory University Goizueta Business School for fulfilling our requests for CRSP data. Errors and omissions are attributable solely to the authors.

REFERENCES

Akerlof, G. A. (1970). The market for 'lemons': Quality uncertainty and the market mechanism. *Quarterly Journal of Economics*, *84*, 488–500.

Allen, F. (1984). Reputation and product quality. *Rand Journal of Economics*, *15*, 311–327.

Benjamin, D. K., & Mitchell, M. L. Commitment and consumer sovereignty: Classic evidence from the real thing, undated manuscript.

Blacconiere, W. G., & Patten, D. M. (1994). Environmental disclosures, regulatory costs, and changes in firm value. *Journal of Accounting and Economics*, *18*, 357–377.

Bolton, G. E. (1991). A comparative model of bargaining: Theory and evidence. *American Economic Review*, *81*, 1096–1136.

Bosch, J., & Lee, I. (1994). Wealth effects of Food and Drug Administration (FDA) decisions. *Managerial and Decision Economics*, *15*, 589–599.

Camerer, C., & Thaler, R. H. (1995). Anomalies: Ultimatums, dictators and manners. *Journal of Economic Perspectives*, *9*, 209–219.

Chalk, A. J. (1987). Market forces and commercial aircraft safety. *Journal of Industrial Economics*, *36*, 61–81.

Chauvin, K. W., & Guthrie, J. P. (1994). Labor market reputation and the value of the firm. *Managerial and Decision Economics*, *15*, 543–552.

Cloninger, D. O., Skantz, T. R., & Strickland, T. H. (1988). Price fixing and legal sanctions: The stockholder-enrichment motive. *Antitrust Law and Economics Review*, 17–24.

Davidson, W. N. III, Worrell, D., & Cheng, L. T. W. (1994). The effectiveness of OSHA penalties: A stock-market-based test. *Industrial Relations*, *33*, 283–296.

Dowdell, T. D., Govindaraj, S., & Jain, P. C. (1992). The Tylenol incident, ensuing regulation, and stock prices. *Journal of Financial and Quantitative Analysis*, *27*, 283–301.

Fry, C. L., & Lee, I. (1989). OSHA sanctions and the value of the firm. *The Financial Review*, *24*, 599–610.

Hamilton, J. T. (1995). Pollution as news: Media and stock market reactions to the Toxics Release Inventory data. *Journal of Environmental Economics and Management*, *28*, 98–113.

Hanka, G. (November 25, 1992). Does Wall Street want firms to harm their neighbors and employees? manuscript.

Harper, R. K., & Adams, S. C. (1996). CERCLA and deep pockets: Market response to the Superfund program. *Contemporary Economic Policy*, *14*, 107–115.

Hayes, A. S., & Pereira, J. (November 8, 1990). Facing a boycott, many companies bend. *The Wall Street Journal*, B1 & B7.

Hersch, J. (1991). Equal Employment Opportunity law and firm profitability. *The Journal of Human Resources*, *26*, 139–153.

Hoffer, G. E., Pruitt, S. W., & Reilly, R. J. (1988). The impact of product recalls on the wealth of sellers: A reexamination. *Journal of Political Economy*, *96*, 663–670.

Ippolito, R. A. (1992). Consumer reaction to measures of poor quality: Evidence from the mutual fund industry. *Journal of Law and Economics*, *35*, 45–70.

Jarrell, G., & Peltzman, S. (1985). The impact of product recalls on the wealth of sellers. *Journal of Political Economy*, *93*, 512–536.

Kahneman, D., Knetsch, J. L., & Thaler, R. H. (1986). Fairness and the assumptions of economics. *Journal of Business*, *59*, S285–S300.

Karpoff, J., & Lott, J. R. Jr. (1993). The reputational penalty firms bear from committing criminal fraud. *Journal of Law and Economics*, *36*, 757–802.

Karpoff, J. M., Lott, J. R. Jr., & Rankine, G. (October 23, 1998). Environmental violations, legal penalties, and reputation costs. *Social Science Research Network*, working paper.

Kihlstrom, R. E., & Riordan, M. H. (1984). Advertising as a signal, *Journal of Political Economy*, *92*, 427–450.

Klein, B., & Leffler, K. B. (1981). The role of market forces in assuring contractual performance, *Journal of Political Economy*, *89*, 615–641.

Klein, B., Crawford, R. G., & Alchian, A. A. (1978). Vertical integration, appropriable rents, and the competitive contracting process. *Journal of Law and Economics*, *21*, 297–326.

Kreps, D. M., & Wilson, R. (1982). Reputation and imperfect information. *Journal of Economic Theory*, *27*, 253–279.

Laplante, B., & Lanoie, P. (1994). The market response to environmental incidents in Canada: A theoretical analysis. *Southern Economic Journal*, *60*, 657–672.

Mathios, A., & Plummer, M. (1989). The regulation of advertising by the Federal Trade Commission: Capital market effects. *Research in Law and Economics*, *12*, edited by Richard O. Zerbe, Jr., JAI Press, Greenwich, Connecticut, 77–93.

Milgrom, P., & Roberts, J. (1986). Price and advertising signals of product quality. *Journal of Political Economy*, *94*, 796–821.

Mitchell, M. L., & Maloney, M. T. (1989). Crisis in the cockpit? The role of market forces in promoting air travel safety. *Journal of Law and Economics*, *32*, 329–356.

Mitchell, M. L. (1989). The impact of external parties on brand-name capital: The 1982 Tylenol poisonings and subsequent cases. *Economic Inquiry*, *27*, 601–618.

Muoghalu, M. I., Robison, H. D., & Glascock, J. L. (1990). Hazardous waste lawsuits, stockholder returns, and deterrence. *Southern Economic Journal*, *57*, 357–370.

Navarro, P. (1988). Why do corporations give to charity? *Journal of Business*, *61*, 65–93.

Nelson, P. (1974). Advertising as information. *Journal of Political Economy*, *82*, 729–754.

Nelson, P. (1970). Information and consumer behavior. *Journal of Political Economy*, *78*, 311–329.

Peltzman, S. (1981). The Effects of FTC advertising regulation. *Journal of Law and Economics*, *24*, 403–448.

Rabin, M. (1993). Incorporating fairness into game theory and economics. *American Economic Review*, *83*, 1281–1302.

Reichert, A. K., Lockett, M., & Rao, R. P. (1996). The impact of illegal business practice on shareholder returns. *The Financial Review*, *31*, 67–85.

Rogers, T. J. (October 1996). Profits vs. PC. *Reason*, *28*, October 1996, 36–43.

Rothchild, J. (May 13, 1996). Why I invest with sinners. *Fortune*, 197.

Rubin, P. H., Murphy, R. D., & Jarrell, G. (1988). Risky products, risky stocks. *Regulation*, 35–39.

Schwartz, R. A. (1968). Corporate philanthropic contributions. *The Journal of Finance*, *23*, 479–497.

Schwert, G. W. (1981). Using financial data to measure effects of regulation. *Journal of Law and Economics*, *24*, 121–158.

Shapiro, C. (1982). Consumer information, product quality, and seller reputation. *Bell Journal of Economics*, *13*, 20–35.

Shapiro, C. (1983). Premiums for high quality products as returns to reputations. *The Quarterly Journal of Economics*, *98*, 659–679.

Viscusi, W. K., & Hersch, J. (1990). The market response to product safety litigation. *Journal of Regulatory Economics*, *2*, 215–230.

CORPORATE BANKRUPTCY, PRIVATE CREDITORS, AND THE MARKET FOR CORPORATE CONTROL

Myron B. Slovin, Marie E. Sushka and Edward R. Waller

ABSTRACT

We assess valuation effects on creditors holding the private debt of client firms that file for Chapter 11 bankruptcy, and disaggregate the results by type of private creditor, form of bankruptcy resolution, and subsequent control bid activity. Among private creditors holding unsecured claims, losses are consistently more severe for trade creditors than for bank creditors. These losses are mitigated for cases in which petitioning firms subsequently reorganize and become targets of control bids during the Chapter 11 process. Banks holding secured claims sustain zero excess returns regardless of the type of bankruptcy resolution or subsequent control activity. Our results support the view that bank loans are effectively senior debt and that banks have a key role in continuation/liquidation decisions. Moreover, we find that control bids for firms in Chapter 11 are common and generate significant gains to equityholders of firms in Chapter 11 and to bidders, so that corporate control activity is an important element in the bankruptcy process.

Advances in Financial Economics, Volume 6, pages 183–208.
Copyright © 2001 by Elsevier Science B.V.
All rights of reproduction in any form reserved.
ISBN: 0-7623-0713-7

I. INTRODUCTION

We analyze the valuation effects of Chapter 11 bankruptcy filings by petitioning firms, and of subsequent events during bankruptcy proceedings, focusing on the two major types of private debtholders, commercial banks and trade creditors. Our results provide perspective on the bankruptcy process and on recent finance literature that views banks as performing a key role in corporate financing as a result of their influence on continuation/liquidation decisions and the reorganization of poorly performing firms. Studies of financial structure, such as Aghion and Bolton (1992) and Dewatripont and Tirole (1994), argue that it is optimal to monitor and discipline managers through the use of incentive-based compensation contracts and the issuance of debt instruments that can force the transfer of control from shareholders to debtholders in response to a firm's poor financial performance. As a result, corporate managers encounter little interference from creditors when the firm is performing well, but suffer a loss of control to creditors when the firm fails to repay debt and/or violates loan covenants. From this perspective, short-term debt, rather than long-term debt, increases the sensitivity of the firm's borrowing costs to new information and allows lenders to exercise control over firms in financial difficulty. Diamond (1993) argues that short-term debt is concentrated in banks because these institutions possess private, as well as public, information and have specific expertise about a borrowing firm's management and operations. As a result, banks are in a strategic position to monitor and influence a financially distressed firm's activities and to make effective liquidation/continuation decisions. Other lenders that lack such information and expertise hold long-term debt as dispersed bondholders. Diamond concludes that bank loans, although short-term, are appropriately senior to other forms of debt.

The ability of a firm to tap the financial market and access short-term debt on a continuous basis, normally limits a bank's control over a borrower. However, when information emerges about a firm's inability to refinance its short-term debt, the firm's lending bank can take actions that directly affect the firm's operations. Given the value of bank-firm relationships, a lending bank will refinance a client firm for which it has favorable private information, but if the firm's prospects are deleterious and the bank's loans are sufficiently senior, either explicitly or implicitly, the bank has little incentive to compromise with the firm. Given adverse selection, the withdrawal of bank financing often leads a financially distressed firm to petition for Chapter 11 protection. From the perspective of Diamond's analysis, banks are sophisticated, cohesive lenders so bank claims should be a central element in explaining the relationship between the market's evaluation of a distressed firm's future prospects and the continuation/liquidation decision that is the focus of Chapter 11 bankruptcy.

We analyze valuation effects at announcements of Chapter 11 filings by operating firms listed on NYSE or AMEX over the period 1980 through 1989. Our results for the valuation effects on petitioning firms are consistent with prior studies by Aharony, Jones, and Swary (1980), Clark and Weinstein (1983), Morse and Shaw (1988), Bradley and Rosenzweig (1992), and Lang and Stulz (1992). However, we also find negative excess returns to trade creditors and banks that hold unsecured private claims of a firm that files for Chapter 11. Negative returns at filing announcements for private creditors, and for equityholders of petitioning firms, are more deleterious for events in which subsequent Chapter 11 resolution is via transfer to Chapter 7 liquidation than for events in which petitioning firms exit through reorganization. These results imply that the market's response to news of a Chapter 11 filing conveys incremental information about expected future cash flows and the petitioning firm's prospects as an ongoing concern.

Within the category of unsecured private creditors, valuation effects are significantly more negative for trade creditors than for bank creditors. Moreover, trade creditor losses are severe relative to the size of their claims, which suggests that a client firm's bankruptcy entails the loss of valuable customer-specific investments and/or future sales for trade creditors. Secured bank creditors, however, earn zero excess returns regardless of bankruptcy outcome, indicating that the market expects the bankruptcy process to provide adequate protection, in economic terms, to holders of secured loans. Negative excess returns to unsecured creditors and normal returns to secured bank creditors imply that the market does not expect the bankruptcy process to distribute wealth away from secured bank creditors to unsecured creditors or to equityholders of petitioning firms. These results suggest that banks holding secured claims against client firms about which they have negative private information have little incentive to compromise with financially distressed borrowers, a result consistent with Diamond's prediction of the effective seniority of short-term bank loans in corporate financing.

Banks engage in various activities during the period that a client firm is in Chapter 11 bankruptcy, including the ongoing monitoring of the managers and activities of the petitioning firm, serving in a dominant position on creditor committees, participating in decisions about the possible termination of managers of the firm or the appointment of a trustee, voting to approve or veto reorganization plans, and deciding whether to retain their claims or sell them to third party strategic investors or other potential bidders for the petitioning firm. Although the Bankruptcy Code contains impediments to outside agents seeking to acquire a firm in Chapter 11, we find that control bids for these firms occur frequently, with more than half of subsequently reorganized firms receiving a control bid during bankruptcy proceedings. At Chapter 11 filing announcements, share price effects on petitioning firms that become targets of control bids, and on their unsecured private creditors, are significantly less deleterious than for cases in which petitioning firms do not receive control bids. Thus, Chapter 11 outcomes and returns to equityholders of petitioning firms and private creditors at filing announcements are correlated. This result suggests that the aggregation of information through market prices about firms filing Chapter 11 and the activities of participants in the bankruptcy process are partially revealing of this information.

We find that equityholders of petitioning firms earn significantly positive returns at the announcement of control bids during bankruptcy proceedings. Almost all of these control bids lead to a change in control. However, if there is no subsequent change in control, gains to equityholders from bid announcements dissipate within three months. Contrary to the negative returns for bidding firms documented in studies of the market for corporate control (e.g. Bradley, Desai & Kim, 1988; Byrd & Hickman, 1992), we find significantly positive returns to bidders for firms in Chapter 11, suggesting that these bidding firm gains are due to restrictions on the bidding process embedded in the Bankruptcy Code. Returns to private creditors at control bid announcements are small and not statistically significant, a result that may reflect changes in the ownership structure of the private debt due to post-petition trading of claims.

The remainder of the chapter is organized as follows. In Section II we discuss the effects of a corporation's Chapter 11 filing and subsequent bankruptcy proceedings on its private creditors, specifically banks and trade creditors. Section III details data and methodology. Empirical results are reported in Section IV, and conclusions are found in Section V.

II. CHAPTER 11 FILINGS BY CORPORATIONS AND THE ROLE OF PRIVATE CREDITORS

Diamond (1993) contends that banks effectively monitor client firms, offer implicit promises of future funding as a central element of lending relationships, and over time acquire valuable private information about interim corporate cash flows. Bank expertise facilitates decisions about whether to support financially distressed client firms in return for greater claims over future corporate earnings. Since bank loans are short-term in maturity, but ongoing in a relationship-based sense, the terms of bank financing are sensitive to new information about interim corporate cash flows. As a result, banks influence whether a borrower continues as an ongoing concern. In particular, bank loan contracts include material adverse change clauses that allow a bank to withdraw its lending or change the terms of a loan, such as by requiring security interests, as deleterious conditions develop at the client firm. When a loan is canceled or the renewal or restructuring of a loan is denied, a bank effectively bars a firm's access to continuation of needed liquidity, a decision that can precipitate a Chapter 11 filing by the firm.

Diamond's hypothesis predicts that firms in financial distress typically have capital structures containing substantial amounts of short-term bank debt that effectively ranks senior in priority. Relative to trade creditors or public debtholders, banks have substantial asymmetric information about borrowers, hold concentrated claims, and develop expertise in dealing with firms in financial distress. Thus, a bank's unwillingness to support a firm under financial pressure explains, at least in part, why there are large negative returns to petitioning firms at the time of Chapter 11 filing announcements even when there has been prior deleterious news about these firms and a prior decline in their stock prices. Moreover, banks have substantial influence over the outcome of the bankruptcy process because of their voting power, the potential to sell their claims to strategic investors, their dominant role on creditor committees vis-à-vis dispersed debtholders, and their ability to pressure a petitioning firm's management due to their credibility in bankruptcy court as sophisticated quasi-insiders. This implies that banks should be effective in furthering their own interests during a client firm's bankruptcy.

Outcomes of Chapter 11 filings can be grouped into three categories. One, a stand-alone or self-reorganization entails negotiations among claimants about a plan that establishes the financial ownership and managerial structure of the emergent firm, that meets the rules prescribed by the Bankruptcy Code, and that is acceptable to the bankruptcy judge. Two, a reorganization can be based on an acquisition by an outside party (i.e. a corporate control transaction) that is agreed to by claimants, subject to approval by the bankruptcy judge. Three, there can be a transfer of a Chapter 11 petition to a Chapter 7 proceeding, in which a court appointed trustee terminates the petitioning firm's operations and sells the firm's assets, with cash proceeds distributed to claimants according to an explicit, nonwaivable priority rule specified in the Bankruptcy Code.

Absolute priority in bankruptcy law is a hierarchical rule that details the ordering of classes of claimants for reimbursement in a bankruptcy resolution and provides that junior claimholders receive nothing until more senior claimants are paid in full. In practice, equityholders and managers of petitioning firms can impose costs on creditors by delaying exit from bankruptcy, given that the Code accords the petitioner an automatic stay on interest payments as well as an exclusivity period to develop a reorganization plan. As a result, creditors may be induced to transfer some of the value of their claims to equityholders of the petitioning firm to avoid or reduce the costs of delay in agreeing to a Chapter 11 resolution. Although secured creditors technically have the right to repossess and sell assets for which they have an enforceable legal lien or to receive cash-equivalents before the priority rule is applied, the Bankruptcy Code stays these rights to preserve prospects for reorganization. The Code requires that secured claims be given adequate protection throughout the bankruptcy process, but no valuation standard is specified in the Code so bankruptcy judges have broad powers to determine adequate protection on a case-by-case basis. This induces a risk for secured creditors since courts may condone petitioning firm actions that are detrimental to collateral value, or the courts may undervalue collateral in post-petition settlements.

The economic question of whether the market expects secured creditors to be compensated for the time value of their claims, and the risk of deterioration of secured assets during bankruptcy proceedings, is a controversial subject in reorganization law (Baird & Jackson, 1984) and raises issues about transfers of wealth between claimant groups. When a Chapter 11 filing culminates in liquidation, secured creditors reclaim their collateral before disbursements are made to other creditor classes so wealth transfers from secured creditors to other claimant classes do not occur. Reorganizations, however, often require secured creditors to accept new equity, deferred cash payments, or illiquid securities in settlement of original claims. Although the face values of the new claims and original claims may be equal, in practice there is often no market for the new claims or they trade at a discount from stated value. This practice can effectively redistribute resources from secured creditors toward junior and residual claimants. Jackson and Scott (1989) argue that bankruptcy court practices, and the Bankruptcy Code, foster such wealth redistribution due to the explicit requirement that a settlement must be endorsed by a majority of those voting, representing at least two-thirds of the principal amounts for each class of creditors. Thus, secured creditors may be induced to accept the new claims to hasten bankruptcy resolution, implicitly agreeing to some redistribution of wealth.

Expectations about bankruptcy proceedings have implications for the market for secured debt. If secured creditors expect to be treated poorly under bankruptcy rules and practice, ex ante the cost of collateralized financing will rise so borrowers will not benefit from greater specialization in bearing risk (Jensen, 1991). However, if secured creditors expect to receive adequate protection, this potential inefficiency is avoided. We examine the issue of wealth transfer by assessing valuation effects on private creditor firms at the time a client borrower files for Chapter 11. If the market expects the bankruptcy process to devalue secured claims, thereby transferring wealth to junior claimants, there should be negative valuation effects on secured creditors at filing announcements by client firms. If the market expects the bankruptcy process to provide adequate protection, there should be no significant share price effects on secured creditors.

There are additional characteristics of private creditors that may influence valuation effects at Chapter 11 filing announcements by client firms. Diamond (1984) argues that banks have a comparative advantage in gathering and processing private information and recontracting with client firms. Rajan (1992) and Diamond (1993) conclude that bank access to private information generated through confidential bank-firm relationships, gives banks considerable influence over borrowers. Since bank portfolios typically contain some loans to firms in poor financial condition, banks invest in developing expertise in post-default collection activity. From this perspective, banks have a comparative advantage at dealing with court-supervised bankruptcy proceedings and making informed decisions about the reorganization or liquidation of a petitioning firm. For less sophisticated creditors, the transactions costs of participating in a bankruptcy proceeding are often prohibitive. From this perspective, valuation effects of Chapter 11 filings should be less deleterious for the firm's bank creditors relative to other creditor classes.

Trade creditors are typically unsecured private creditors that supply inputs to customer firms and have explicit and/or implicit contracts about future commercial transactions. When a customer files Chapter 11, trade creditors can lose future sales and non-salvageable customer-specific capital in addition to sustaining losses on outstanding debt claims, which potentially can lead to financial distress for these suppliers. Moreover, compared to bank claims, trade creditor claims are often less concentrated, which suggests that valuation effects at customer bankruptcy announcements should be more deleterious for unsecured trade creditors than for unsecured bank creditors. Alternatively, if trade credit is a self-selection device that allows suppliers to reduce investments in customer-specific capital and diversify their clientele (Smith, 1987), unfavorable share price effects should be mitigated at customer Chapter 11 filings. This view assumes that trade creditors are able to quickly limit their investments in failing firms and shift their activities toward healthier firms in a related line of business.

Previous market-based studies report sizable negative returns to shareholders of petitioning firms at Chapter 11 announcements and negative returns to public bondholders. However, there is no market-based analysis of the effects of a Chapter 11 filing on bank creditors or trade creditors. Our study develops such evidence given that the private creditors in our analysis are publicly traded entities.

Although bankruptcy affords debtors-in-possession, that is, petitioning firms, considerable benefits vis-à-vis their creditors, including an automatic stay on interest payments and an exclusivity period to develop a reorganization plan, there are mechanisms that can limit these strictures of the bankruptcy process. One mechanism is that banks typically have influential positions on creditor committees, given their sizable claims and the dispersion of other debtholders. Due to knowledge gained from interaction with petitioning client firms, banks have considerable influence on the reorganization process and on managerial decisions during bankruptcy proceedings. For example, the bankruptcy court consults with creditor committees about the petitioning firm's investment plans and managerial retention and compensation plans. Bank claims are often large and concentrated, giving banks the voting power to block a management-proposed reorganization plan and to reduce managerial bargaining power.

Another mechanism is the buying and selling of claims on bankrupt firms, a sizeable component of which is bank debt. Private creditors who are not interested in being involved in Chapter 11 proceedings, or who prefer alternative investment opportunities, can influence bankruptcy outcomes by selling their claims to post-petition investors. The process of selling such claims entails adverse selection bias as a consequence of asymmetric information problems intrinsic to firms in financial distress, but bankruptcy filings trigger a substantial flow of corporate information in a public forum that may lessen these asymmetric information problems. As a result, some creditors may have an incentive to sell their claims, especially to sophisticated strategic investors who gather information about bankrupt firms and assess what proceeds will be realized through a stand-alone reorganization or a transfer of control to an acquirer. Post-petition investors with resources and experience in working out bankruptcies can also provide elements of monitoring and control, enhancing the value of petitioning firms as well as the value of creditor claims, and increasing the value of new securities to be distributed to claimants in a reorganization. Claims trading increases creditor concentration while enhancing the role of sophisticated post-petition investors who can facilitate and influence bankruptcy resolution. In particular, concentrated claimholders and strategic post-petition investors can facilitate activity in the market for corporate control of firms in bankruptcy. Moreover, a third party bid provides a market-based benchmark of the value of the petitioning firm against which competing reorganization plans, including management's plan, can be compared.

Since the 1978 revision of the Bankruptcy Code, outside investors, both individual and corporate, have entered the bankruptcy reorganization process (Hotchkiss & Mooradian, 1997). Relative to judicially administered reorganizations, the sale of a bankrupt firm to a third party is likely to maximize value because it minimizes valuation errors and transactions costs (Baird, 1986). In many cases, the post-petition purchase of bank claims provides the basis for outside bidders to acquire control of a bankrupt firm. An outside agent can make the content or outline of a control bid public without violating petitioner exclusivity per se, since a bid is not a formal reorganization plan. The bid establishes a credible estimate of petitioning firm value based on information collected by the bidder who has an incentive to accurately assess the current condition of the firm. This may induce the management of a Chapter 11 firm to submit a reorganization plan that includes the bidder, or that meets or exceeds the value of the outside bid. This perspective suggests that even a failed control bid generates a market-based standard of firm value that creditors, judges, and other participants can consider in assessing reorganization plans. An implication of this view is that if takeover bids generate gains to petitioning firm shareholders, these gains should persist even if a change in control does not subsequently succeed. However, if expected gains in productivity or improvements in management associated with a bid are contingent on an actual change in control, any gains in value at the bid announcement should dissipate if no control change occurs.

The Bankruptcy Code does not provide specific guidance about the treatment of corporate control activities involving Chapter 11 firms. Bidders face significant uncertainty and ambiguity since bankruptcy judges have broad powers to influence the reorganization process, including taking actions that block a bidder from gaining control and deciding that other proposals, including the reorganization plan submitted by the petitioning firm's management, would be fairer to claimants. In addition, post-petition purchases of creditor claims by outside agents can jeopardize the ability of the petitioning firm to maintain net operating loss carry forwards, which are among the most valuable assets of a bankrupt firm (Fortgang & Mayer, 1990). Overall, actions by outside agents to gain control of a bankrupt firm take place in a more uncertain and legally cumbersome setting compared to the environment in which typical corporate control activity occurs.

Studies of the market for corporate control, typified by Bradley, Desai, and Kim (1988) and Byrd and Hickman (1992), analyze effects on target firms that are not in distress and report that bidding firms during the 1980s sustain significantly negative returns at control bid announcements, suggestive of overbidding. Clark and Ofek (1994) report negative returns to bidders for firms that are in financial distress, but not in bankruptcy. French and McCormick's (1984) auction model predicts that restrictions on bidders negatively affect selling (target) firm price and transfer wealth to bidders. Consistent with this model, James and Wier (1987) and Giliberto and Varaiya (1989) find positive and significant returns to acquirers of insolvent banks sold through FDIC auctions, which they ascribe to regulatory restrictions on bidder participation, since studies of conventional acquisitions of solvent banks typically report negative returns to bidders (Cornett & Tehranian, 1992; Cornett & De, 1991; Sushka & Bendeck, 1988). We assess the extent of takeover activity in bankruptcy proceedings, and provide evidence about share price responses of targeted bankrupt firms, bidder firms, and private creditors. If restrictions intrinsic to the Bankruptcy Code constrain bidding for bankrupt firms, there should be positive returns to bidders and reduced gains to targets, indicating a transfer of wealth from targets to bidders, relative to an unrestricted control contest. Examination of the extent of bidding for firms in Chapter 11 also provides insight about whether the valuation effects of filing announcements on petitioning firms and their private creditors reflect expectations about subsequent control bid activity.

III. DATA AND METHODOLOGY

Our sample consists of 106 announcements of Chapter 11 filings by NYSE or AMEX listed industrial firms that are trading at the date of the bankruptcy filing. Of the 106 petitioning firms, there are 81 firms that have publicly traded creditor firms holding their private debt. The sample period is 1980 through 1989 which provides a suitable post-event time period to examine subsequent developments and resolutions of Chapter 11 petitions. The filing announcement dates are obtained from the *Wall Street Journal*. The distribution of events by year is reported in Panel A of Table 1. By 1998, 81 firms exit Chapter 11 through reorganization plans and 20 firms transfer to Chapter 7 liquidation. Petitioning firms are widely dispersed across industries, representing 75 different 4-digit Standard Industrial Classification (SIC) codes. Grouped by type of resolution, there are no apparent differences in the pattern of SIC codes.

Table 2 contains descriptive statistics for ex ante financial characteristics of the petitioning firms based on data that most closely predate the Chapter 11 filing. As reported in Panel A, banks have large, concentrated holdings in bankrupt firms, which supports the view that banks play a major role in the continuation/liquidation decision. Firms in Chapter 11 that subsequently transfer to Chapter 7 liquidation have larger bank debt ratios and a smaller number of debt contracts, which are generally secured, relative to firms that subsequently reorganize. In addition, the book value of total assets is smaller for petitioning firms that liquidate rather than reorganize. Firms spend considerable time in legal proceedings under Chapter 11 regardless of whether the outcome is reorganization (median of 24 months) or liquidation (median of 29 months). Panel B contains descriptive statistics for the set of firms that eventually reorganize, disaggregated by subsequent control bid status. We find no significant

Table 1. Frequency Distribution of Chapter 11 Filings and Firm Size of Private Creditors.[1]

Panel A: Firms Filing Chapter 11

Year (1)	Total (2)	With Creditors (3)
1989	4	1
1988	13	10
1987	11	9
1986	13	12
1985	15	11
1984	12	11
1983	6	4
1982	15	11
1981	9	7
1980	8	5
Total	106	81

Panel B: Median Market Value of Private Creditors (Million $)

	Full Sample (1)	Banking Institutions (2)	Trade Creditors (3)
Total Creditors	1157.4	1153.2	1314.9
(N)	(266)	(233)	(33)
Secured Creditors	1632.2	1632.2	–
(N)	(76)	(76)	–
Unsecured Creditors	1049.3	1045.9	1314.9
(N)	(190)	(157)	(33)

Notes: [1] Frequency distribution of Chapter 11 filing announcements by industrial firms listed on NYSE/AMEX and median firm size (calculated as price multiplied by shares outstanding seven trading days prior to the bankruptcy announcement, million $) of private creditors listed on NYSE/AMEX/NASDAQ over the period 1980 through 1989; N is the sample size.

differences in the characteristics of these two sets of firms that explain, ex ante, whether a control bid will be received.

We examine survivorship of petitioning firm top management, defined as CEO, Chairman of the Board, or President, during Chapter 11 proceedings.

Table 2. Ex Ante Characteristics of Firms Filing Chapter 11.[1]

Characteristics	Mean (1)	Median (2)	Mean (3)	Median (4)	Difference in Means (5)	Difference in Medians (6)
Panel A: By Type of Resolution	Reorganizations N=81		Liquidations N=20		(p-value)	(p-value)
Market Value/Total Liabilities	0.50	0.25	0.45	0.19	(0.95)	(0.66)
Long-term Debt/Book Value of Assets	0.32	0.28	0.40	0.37	(0.20)	(0.16)
Total Liabilities/Book Value of Assets	0.72	0.71	0.82	0.74	(0.12)	(0.25)
Bank Debt/Total Liabilities	0.17	0.12	0.30	0.24	(0.10)	(0.01)
Public Debt/Total Liabilities	0.11	0.02	0.11	0.00	(1.00)	(0.42)
Secured Debt/Total Liabilities	0.19	0.06	0.21	0.15	(0.49)	(0.20)
Convertible Debt/Total Liabilities	0.04	0.00	0.05	0.00	(0.73)	(0.72)
Number of Debt Contracts	6.52	6.00	4.30	4.00	(0.01)	(0.05)
Book Value of Total Assets (million $)	870.31	168.93	72.56	48.18	(0.07)	(0.00)
Time in Chapter 11 (months)	25.72	23.60	27.51	28.65	(0.69)	(0.74)
Panel B: By Control Bid Status	Control Bid N=47		No Control Bid N=34		(p-value)	(p-value)
Market Value/Total Liabilities	0.45	0.22	0.57	0.29	(0.54)	(0.73)
Long-term Debt/Book Value of Assets	0.30	0.28	0.33	0.34	(0.50)	(0.41)
Total Liabilities/Book Value of Assets	0.71	0.70	0.73	0.75	(0.58)	(0.62)
Bank Debt/Total Liabilities	0.16	0.11	0.18	0.17	(0.65)	(0.27)
Public Debt/Total Liabilities	0.10	0.00	0.12	0.06	(0.54)	(0.30)
Secured Debt/Total Liabilities	0.22	0.05	0.15	0.07	(0.38)	(0.97)
Convertible Debt/Total Liabilities	0.04	0.00	0.04	0.00	(1.00)	(0.30)
Number of Debt Contracts	5.85	5.00	7.44	7.00	(0.16)	(0.08)
Book Value of Total Assets (million $)	1035.75	171.32	641.62	143.36	(0.61)	(0.76)
Time in Chapter 11 (months)	24.80	22.80	26.98	25.60	(0.52)	(0.78)

Notes: [1]Selected firm and debt characteristics of 81 Chapter 11 petitioning firms that subsequently reorganize and 20 firms that transfer their Chapter 11 petition to Chapter 7 liquidation (Panel A), and 47 reorganized firms that receive control bids during bankruptcy and 34 reorganized firms that do not receive control bids during bankruptcy (Panel B). All petitioning firms are traded on NYSE/AMEX. The sample period is 1980 through 1989 and the data are those that most closely predate the Chapter 11 filing. Balance sheet figures are obtained from Moody's manuals. For liquidations, the length of the period in Chapter 11 is defined as the time from the Chapter 11 filing until the date the filing is transferred to Chapter 7.

Changes in management reported in the *Wall Street Journal* indicate that turnover is generally involuntary, typically with indications of pressure from creditor committees. The primary reason given for turnover is the need to establish a team of executives that can better manage the bankruptcy process and gain the confidence of creditor committees and the bankruptcy judge. Management is automatically removed when a Chapter 11 petition is transferred to Chapter 7 since operations are terminated and a trustee is appointed. In our sample, we find that a majority of the firms that reorganize sustain management changes, which generally occur early in the bankruptcy process. For the 81 reorganizations, top management is replaced at least once for 52 firms, typically as the result of creditor pressure, a finding that is consistent with the view that banks play an important monitoring role throughout the bankruptcy process. Turnover is clearly voluntary in only two cases. Management is replaced by a trustee in four reorganization cases. Overall, our results our similar to the turnover rates reported by Hotchkiss (1995), 70%, and Betker (1995), 75%.

The set of private creditors of the sample of 106 petitioning firms is obtained from articles in the *Wall Street Journal*, the *New York Times*, the Dow Jones News Wire, and Lexis-Nexis. We retain in the sample the private creditors that are publicly traded on NYSE, AMEX, or NASDAQ. The Bankruptcy Code requires the petitioning firm to disclose its major creditors as part of the Chapter 11 filing, which is typically the first explicit public information about creditor identity by class, security status, and face value of claims. As reported in Panel B of Table 1, our sample consists of 266 publicly traded private creditors associated with 81 of the 106 firms filing Chapter 11. There are 76 creditors with secured claims, all of which are banks, and 190 unsecured creditors. Among the 266 private creditors, there are 233 banks, which make up 100% of the secured creditors and 83% of unsecured creditors. There are 33 trade creditors, all unsecured. The relatively small sample of trade creditors reflects the fact that many trade creditors are private firms or are not traded on NYSE, AMEX, or NASDAQ. Median market values (measured as shares outstanding multiplied by price seven days prior to the event) of secured creditors are greater than unsecured creditors, but the median market value of trade creditors exceeds that of bank creditors, which likely reflects the highly leveraged capital structure of banks relative to nonfinancial firms.

Table 3 reports the mean (median) face value of private claims of creditors and the relevant credit exposure the claims represent. Credit exposure is measured as the ratio of the face value of each claim to the relevant creditor's market capitalization. The average face value of claims is $28.1 million (median is $12.6 million), representing a credit exposure of 4.6% (median is 1.4%), although these figures vary by type of creditor and security status. The face

Table 3. Face Value and Credit Exposure of Private Creditors of Firms Filing Chapter 11.[1]

	Face Value of Claims (million $)				Credit Exposure (%)			
	Total	Reorganized	Liquidated	Difference in Means (Medians)	Total	Reorganized	Liquidated	Difference in Means (Medians)
	(1)	(2)	(3)	(4)	(5)	(6)	(7)	(8)
Panel A: Creditors by Security Status								
All Creditors	28.1	29.3	17.0	1.97[2]	4.6	4.7	3.4	0.21
[208,191,15]	(12.6)	(12.0)	(10.0)	(0.89)	(1.4)	(1.4)	(0.8)	(0.78)
Secured	45.9	50.0	0.3	3.67[3]	4.8	5.2	1.6	2.45[2]
[70,62,7]	(29.0)	(31.9)	(8.0)	(2.20)[2]	(1.9)	(2.1)	(0.4)	(1.92)[2]
Unsecured	19.1	19.3	19.1	0.11	4.5	4.5	5.0	0.06
[138,129,8]	(10.0)	(10.0)	(10.0)	(0.26)	(1.1)	(1.1)	(1.3)	(0.57)
Difference in Means (Medians)	4.50[3] (4.79)[3]	4.78[3] (5.14)[3]	0.46 (0.41)		0.12 (2.87)[3]	0.20 (3.50)[3]	0.85 (0.98)	
Panel B: By Type of Unsecured Creditor								
Banks	20.2	20.5	19.1	0.03	5.0	5.0	5.0	0.01
[123,114,8]	(10.3)	(10.8)	(10.0)	(0.06)	(1.3)	(1.3)	(1.3)	(0.99)
Trade Creditors	9.0	9.0	—	—	0.8	0.8	—	—
[15,15]	(3.8)	(3.8)	—	—	(0.1)	(0.1)	—	—
Difference in Means (Medians)	2.65[3] (3.50)[3]	2.65[3] (3.50)[3]			1.58 (4.35)[3]	1.58 (4.35)[3]		

Notes: [1] Face value of claims and credit exposure for the total sample and sub-samples formed on the basis of type of bankruptcy resolution, security status, and type of unsecured creditor. The differences in means (medians) for reorganized and liquidated petitioning firms are reported in columns (4) and (8), for secured and unsecured creditors in the last row of Panel A, and for unsecured banking firms and trade creditors in the last row of Panel B; the relevant sample sizes are in brackets. Credit exposure is measured as the ratio (in percent) of the face value of claims to creditor firm market value (calculated as price multiplied by shares outstanding seven trading days prior to the Chapter 11 filing announcement).

[2] Statistical significance at the 5% level.

[3] Statistical significance at the 1% level.

value of creditor claims on petitioning firms that subsequently reorganize is significantly greater than the face value of creditor claims against the firms that liquidate, given a difference in means calculated t-value of 1.97. Consistent with Diamond's hypothesis about the seniority of bank claims, we find that secured bank debt is common and secured bank claims are larger than unsecured bank claims, with the differences in both the mean and median values significant at the 1% level. For 15 trade creditors (each with a different 4-digit SIC code and 13 with different 2-digit SIC codes), the face value of claims and credit exposure ($9.0 million and 0.8%) are significantly smaller, in mean and/or median terms, than for unsecured banks ($20.5 million and 5.0%). We have no information about trade creditor claims for liquidations.

To conduct the event studies we calculate average prediction errors for petitioning firms and for private creditors around the announcement date of Chapter 11 filings, t = 0, using the market model. Market model parameters are estimated using least squares regressions over a 120 day pre-event period, −240 to −121, where the explanatory variable is the relevant CRSP market index, depending on the firm's trading venue. The results are not sensitive to the use of alternative pre-announcement estimation periods. We generate private creditor adjusted returns and pool these creditors into event portfolios, that is, portfolios for each petitioning firm. Thus, returns are adjusted for each creditor firm and then are averaged by event each day to create an event portfolio return. Adjusted returns for each portfolio are averaged over all events for each day in the relevant event period, and cumulated over the number of days in the event interval.

IV. EMPIRICAL RESULTS

In Table 4 column (1), we report the average excess return for petitioning firms measured over a three-day event window, −2, 0, where day 0 is the date of the Chapter 11 filing report in the *Wall Street Journal*. The excess return for the full sample is −29.58% with a t-statistic of −36.83; 93% of returns are negative. This result is similar to the three-day returns to firms filing for bankruptcy as reported by Clark and Weinstein (1983) and Lang and Stulz (1992). For petitioning firms that subsequently reorganize, column (2), excess returns are −24.84% (t-statistic of −30.31) and are significantly less negative than for petitioning firms that liquidate, −39.53% (t-statistic of −16.32), column (3), given a difference in means calculated t-value of 3.38 (p-value = 0.00). These results indicate that market responses to Chapter 11 filing announcements reflect incremental information about bankruptcy outcome. There is no statistical difference in market performance prior to Chapter 11 filings, given that cumulative average daily

Table 4. Three-day Average Excess Returns to Private Creditors and Petitioning Firms at Chapter 11 Filing Announcements.[1]

	Full Sample (1)	Petitioner Reorganized (2)	Petitioner Liquidated (3)
Equity of Petitioner	**−29.58%**	**−24.84%**	**−39.53%**
	(−36.83)[3]	(−30.31)[3]	(−16.32)[3]
	{106}	{81}	{20}
	[0.93]	[0.91]	[1.00]
Creditor Firms Holding Private Debt	**−0.94%**	**−0.80%**	**−2.12%**
	(−3.88)[3]	(−3.03)[3]	(−3.36)[3]
	{81,266}	{66,243}	{13,20}
	[0.70]	[0.68]	[0.90]
Unsecured	**−1.26%**	**−0.95%**	**−3.21%**
	(−4.77)[3]	(−3.34)[3]	(−4.16)[3]
	{67,190}	{56,177}	{10,12}
	[0.75]	[0.71]	[1.00]
Secured	**−0.11%**	**0.02%**	**−0.26%**
	(−0.22)	(0.03)	(−0.21)
	{34,76}	{26,66}	{6,8}
	[0.53]	[0.46]	[0.67]
Unsecured Banks		**−0.67%**	
		(−2.34)[2]	
		{53,145}	
		[0.66]	
Trade Creditors		**−3.89%**	
		(−4.32)[3]	
		{17,32}	
		[0.82]	

Notes: [1] Average prediction errors are calculated using a market model estimated over the pre-event period −240 to −121. Private creditor excess returns are averaged for each bankruptcy event for each day, then averaged over all relevant events, and cumulated over the three days in the event period. The sample period is 1980 through 1989 for filing announcements; resolution is identified through year-end 1998. Liquidated is defined as bankruptcy resolution where the petitioning firm exits Chapter 11 by transferring the filing to Chapter 7. Three-day excess returns are reported in percent, t-statistics are in parentheses, the number of bankruptcy events and associated number of creditors are in braces, and the proportion of returns that are negative is in brackets.

[2] Statistical significance at the 5% level.

[3] Statistical significance at the 1% level.

excess returns over the six months prior to Chapter 11 announcements are −33.81% for firms that subsequently reorganize and −30.11% for firms that subsequently liquidate (not reported in the table). The difference in means calculated t-value is 0.18, indicating equality of pre-bankruptcy share price performance disaggregated by ex post resolution. These results suggest that, although the market may anticipate bankruptcy, the anticipation is only partial since there are large negative valuation effects on petitioning firms at the news of a Chapter 11 filing, and a significant difference in these returns based on ex post bankruptcy outcome.

Private creditors sustain significant losses due to client bankruptcy, given a three-day average excess return of −0.94% with a t-statistic of −3.88; 70% of returns are negative. The two-week pre-event and post-event period returns for private creditors (not reported in the table) are not significantly different from zero. Since there are significantly negative cumulative average returns for equityholders of petitioning firms prior to Chapter 11 filings, normal pre-filing excess returns for private creditors indicate that the market is either unable to identify the creditors and/or unable to learn the extent or status of the creditors' exposure prior to reports required as part of the Chapter 11 filing. The excess returns to private creditors indicate that the median expected loss in the value of the claims of the full sample of private creditors is 42% of the face value of the claims (calculated as the change in shareholder wealth for each creditor divided by the face value of the relevant claim). This expected loss implies an expected recovery rate of 58%.

We find that the market differentiates between classes of private creditors on the basis of security. For unsecured creditors, the three-day average excess return is −1.26% with a t-statistic of −4.77; 75% of returns are negative. For secured bank creditors, the average excess return is effectively zero, −0.11% (t-statistic of −0.22). A difference in means test for the two sets of creditor returns yields a calculated t-value of 2.91 (p-value = 0.00), which indicates that the effect of a Chapter 11 filing on private creditor value is related to whether or not the claim is secured. The median loss in the value of unsecured creditors relative to their claims is 63%, an expected recovery rate of 37%. This expected recovery rate for unsecured creditors is higher than the 25% figure reported by Lopucki and Whitford (1990), in which recoveries are measured as the proportion of the face value of unsecured creditor claims paid at reorganization. Our empirical results imply that there is a median expected recovery rate of 100% for secured bank creditors, indicating that the market does not expect secured bank creditors to sustain losses from client firm bankruptcy. Our evidence is consistent with the view that secured bank creditors are expected to be fully compensated for delays in payment of interest and principal, uncertainty, and other potential costs associated with the bankruptcy process.

The excess return to private creditors of petitioning firms that subsequently reorganize, column (2), is −0.80% (t-statistic of −3.03 with 68% of the returns negative), compared to an excess return of −2.12% (t-statistic of −3.36 with 90% of the returns negative) to private creditors of petitioning firms that liquidate, column (3). Thus, the percentage loss in market value for private creditors of firms that subsequently liquidate is more than twice the loss for private creditors of firms that reorganize, and the difference in means test for these returns generates a calculated t-value of 1.67 (p-value = 0.10). Overall, the returns suggest that the market's response to news of a Chapter 11 filing reflects information about private creditors' ability to recover claims as well as about the petitioning firm's ability to reorganize.

Disaggregating by security status and subsequent bankruptcy outcome, unsecured private creditor excess returns are −0.95% (t-statistic of −3.34) in cases of petitioning firm reorganization, and −3.21% (t-statistic of −4.16) in cases of firms that subsequently liquidate. The difference in these excess returns is statistically significant given a calculated t-value of 2.22 (p-value = 0.04). Our results suggest that at the time of a Chapter 11 filing announcement, the valuation effects on unsecured private creditors of firms that subsequently reorganize are less deleterious than for firms that subsequently liquidate.

For secured bank creditors, excess returns are 0.02% (t-statistic of 0.03) in cases of petitioning firm reorganization, and −0.26% (t-statistic of −0.21) in cases of liquidation. These results indicate that secured bank creditors are expected to receive protection adequate to maintain the present value of their claims, regardless of whether the petitioning firm reorganizes or liquidates. Our evidence is consistent with Diamond's (1993) prediction of the effective seniority of bank lending and indicates that banks with secured claims that have negative private information about a distressed borrower have little incentive to compromise with the client firm to avoid its filing of a Chapter 11 petition and potential liquidation.

For the sample of unsecured private creditors of reorganized petitioning firms, the excess return for banks is −0.67% (t-statistic of −2.34), which is significantly less negative than the return to trade creditors, −3.89% (t-statistic of −4.32), given a difference in means calculated t-value of 1.98 (p-value = 0.05). The median implied loss in the value of unsecured bank claims is 52%, a recovery rate of 48%. The median implied loss in the value of trade creditor claims is an extraordinary 285%, indicating that the decline in the market value of trade creditors at Chapter 11 filings is much greater than the face value of their claims. The severity of these losses likely reflects expectations of lost sales and valuable customer-specific capital as well as the dispersed character of trade creditor claims relative to bank claims. The smaller decline in wealth for banks holding unsecured claims than for trade creditors is consistent with Diamond's model of the banking firm, which predicts that banks have a comparative advantage relative to non-financial (trade) creditors in dealing with the bankruptcy of client firms.

Although there is trading of claims against Chapter 11 firms, the market for these claims is sporadic and private so that no detailed public data are available about this activity. Moreover, since we are unable to access specific court filings, we are unable to directly determine whether each private creditor in the sample retains its claims throughout Chapter 11 proceedings, and if not, at what point in the proceedings the claims are traded and to what type of buyer. However, we can observe control bids for our sample of Chapter 11 firms in public reports. This allows us to indirectly gauge the role of private creditors in the bankruptcy process since either the expectation or the announcement of third party bids may influence the value of these claims against petitioning firms, given that private creditor support is an important facet of the structuring of bids for bankrupt firms by strategic investors or other third party bidders. Moreover, bidders for bankrupt firms often acquire bank claims, which are large and concentrated, a practice that enhances the marketability of private debt held against Chapter 11 firms.

Several issues are relevant in this analysis. One issue is whether control bids for Chapter 11 firms are common events and whether a change in control ensues following a bid. A second issue is the valuation effects of control bids on private creditors as well as on equityholders of petitioning firms. If the identity of private creditors changes after the Chapter 11 filing as a result of claims trading, no valuation effect should be observed at the control bid announcement for the set of private creditors identified at the Chapter 11 filing. If private creditors retain their claims and a control bid enhances target firm value, as is typical for conventional acquisitions, a positive valuation effect should be observed for private creditors at the announcement of a control bid. A third issue is whether a control bid is partially anticipated at the time of the Chapter 11 filing. If a subsequent control bid is partially anticipated, the share price effect at the filing announcement on private creditors and equityholders of petitioning firms should be less unfavorable than at Chapter 11 filings in which no subsequent control bid is expected. Moreover, if no control bid is expected, the claims of private creditors may be difficult to trade so these creditors suffer a loss in liquidity. A fourth issue is that the valuation effects on acquirers at bid announcements can provide evidence as to whether bids for Chapter 11 firms enhance the value of bidders or whether the overbidding that is observed in studies of conventional acquisitions also characterizes bids for control of firms in Chapter 11.

We search the *Wall Street Journal* and other public sources of information about post-petition control bids for the sample of 81 Chapter 11 firms that subsequently reorganize, and find 47 firms (58%) that receive control bids during bankruptcy proceedings. Of the 20 firms that file for Chapter 11 and later transfer the petition to Chapter 7, there is only one (unsuccessful) control bid. According to Dodd (1983), for the population of CRSP firms from 1962 to 1982, 5.7% are targets of tender offers and 9.6% are targets of mergers. Thus, based on our sample, there is a bid for control for a larger percentage of bankrupt firms than for the general population of firms.

The high degree of corporate control activity for firms in Chapter 11 proceedings is important evidence because several provisions of the Bankruptcy Code should influence bidders' willingness to seek control of a Chapter 11 firm relative to conventional corporate control contests. The Code requires the petitioning firm to make extensive disclosures of financial and operating data that often include court testimony by managers. These disclosures provide outside agents with valuable information about the petitioning firm and its outstanding claims and narrow the informational advantage of managers over potential bidders. However, the Code also contains impediments about the extent and timing of bidder activities, provisions that constrain the opportunity for bidding compared to the process of bidding for nonbankrupt firms. Overall, we find that control bids for Chapter 11 firms are common and occur throughout the period of bankruptcy proceedings, results that suggest outside agents have the expertise to circumvent the legal strictures of Chapter 11 and undertake bids for bankrupt firms.

We disaggregate the sample of subsequently reorganized firms based on whether or not there is a control bid. As reported in Table 5, at announcements of Chapter 11 filings the three-day average excess return to 47 firms that subsequently become control targets is −21.41% (t-statistic of −19.30), column (1), compared to an excess return of −35.86% (t-statistic of −31.88) for 34 firms that reorganize without receiving a control bid, column (2). A difference in means calculated t-value is 3.35 (p-value = 0.00), indicating that share price responses to Chapter 11 filings incorporate market expectations about the probability of subsequent control activity. Secured bank creditors earn normal returns, regardless of whether control activity occurs. For unsecured private creditors of petitioning firms that receive bids, the three-day excess return at Chapter 11 filing announcements is −0.38%, not significant given a t-statistic of −1.07. Disaggregating the sample of unsecured creditors, we find trade creditors sustain a loss of −3.42% (t-statistic of −3.24), while the share price response for banks with unsecured claims is effectively zero, −0.10% (t-statistic of −0.27). For unsecured private creditors of petitioning firms that do not receive control bids, the excess return is −1.76% (t-statistic of −5.62). Within this category of unsecured private creditors, the excess returns are −5.01% (t-statistic of −4.04) for trade creditors and −1.54% (t-statistic of −3.61) for banks with unsecured claims. Difference in

Table 5. Three-day Average Excess Returns to Private Creditors and Petitioning Firms at Chapter 11 Filing Announcements and at Control Bid Announcements during Chapter 11 for the Sample of Firms Subsequently Reorganized.[1]

	At Announcement of Chapter 11 Filing		At Announcement of Control Bid
	Receives Control Bid (1)	No Control Bid (2)	(3)
Equity of Petitioning Firm	**−21.41%**	**−35.86%**	**26.10%**
	(−19.30)[3]	(−31.88)[3]	(13.64)[3]
	{47}	{34}	{25}
	[0.92]	[0.94]	[0.12]
Creditor Firms Holding Private Debt			
Unsecured	**−0.38%**	**−1.76%**	**−0.31%**
	(−1.07)	(−5.62)[3]	(−0.66)
	{33,109}	{23,68}	{33,109}
	[0.67]	[0.83]	[0.67]
Secured	**0.14%**	**−0.09%**	**0.03%**
	(0.21)	(−0.08)	(0.03)
	{12,23}	{14,43}	{12,23}
	[0.50]	[0.43]	[0.42]
Unsecured Banks	**−0.10%**	**−1.54%**	**−0.16%**
	(−0.27)	(−3.61)[3]	(−0.35)
	{32,86}	{21,59}	{32,86}
	[0.59]	[0.76]	[0.59]
Trade Creditors	**−3.42%**	**−5.01%**	**−1.03%**
	(−3.24)[3]	(−4.04)[3]	(−0.70)
	{12,23}	{5,9}	{12,23}
	[0.75]	[1.00]	[0.83]
Bidder Firm in Control Offer			**3.03%**
			(2.83)[3]
			{28}
			[0.36]

Notes: [1] Average prediction errors are calculated using a market model estimated over the pre-event period −240 to −121. Private creditor excess returns are averaged for each bankruptcy event for each day, then averaged over all relevant events, and cumulated over the three days in the event period. The sample period is 1980 through 1989 for filing announcements; resolution is identified through year-end 1998. Liquidated is defined as bankruptcy resolution where the petitioning firm exits Chapter 11 by transferring the filing to Chapter 7. Three-day excess returns are reported in percent, t-statistics are in parentheses, the number of bankruptcy events and associated number of creditors are in braces, and the proportion of returns that are negative is in brackets.

[2] Statistical significance at the 5% level.

[3] Statistical significance at the 1% level.

means tests indicate that banks with unsecured claims and trade creditors are each more adversely impacted at Chapter 11 filing announcements by firms that do not receive subsequent control bids compared to filings by firms that become subsequent targets of bids, with trade creditors more adversely affected than banks. Our evidence suggests that expectations about control bids are a significant factor in the market's valuation of private creditors and their client firms at Chapter 11 announcements. Moreover, the valuation effects for trade creditors at Chapter 11 filings are consistently more severe than for bank creditors.

At the announcement of a control bid, there are 25 target firms that are publicly trading and have returns on the CRSP NYSE, AMEX, or NASDAQ files. The three-day average excess return to these petitioning firms at the control bid is 26.10% with a t-statistic of 13.64, column (3). This result suggests that although the market anticipates control bids at the Chapter 11 filing date, this anticipation is only partial since at the control bid announcement, there is a large positive valuation effect on petitioning firm equity. However, the returns to secured and unsecured private creditors at the control bid announcement are small and not statistically significant. The absence of significant excess returns for unsecured private creditors may reflect the prevalence of trading of private claims against firms in Chapter 11, so that the identity of claimholders changes between the Chapter 11 filing and the control bid announcement.

If restrictions intrinsic to the Bankruptcy Code and those imposed by bankruptcy judges exclude or discourage some outside parties from bidding for bankrupt firms or raise the cost of bidding, there should be positive returns to bidders at the expense of targeted bankrupt firms. Contrary to the negative returns to bidders observed in conventional acquisitions, the 28 publicly traded bidders for our sample of firms in Chapter 11 sustain positive excess returns at announcements of control bids. The three-day excess return to bidders is 3.03% (t-statistic of 2.83); 64% of the returns are positive, column (3). Both the pre- and post-event period returns to bidders are small and not statistically significant (not reported in the table). For 15 of the 28 events, both the target and the bidder have equity trading at the bid announcement, which allows us to calculate the distribution of the gains in wealth for these control bids. The statistically significant returns for this group of events are 33.11% for targets and 4.27% for bidders (not reported in the table), with calculated aggregate gains in wealth of $412 million for bidders and $443 million for targets. The $412 million gain in wealth for target firms is a partial recovery of the $1,220 million losses incurred by petitioning firm equityholders at the Chapter 11 filing announcements. Our evidence supports the French and McCormick (1984) model which predicts that restrictions on bidders reduce gains to target firms and transfer wealth to bidders relative to unrestricted corporate control contests.

We document the extent of subsequent changes in control for firms in Chapter 11 proceedings, by collecting information for each firm for each year after the filing through year-end 1998 from the *Wall Street Journal* index and articles, Dow Jones News Wire reports, other business publications, and relevant corporate reports. We group reorganized firms into three categories: (1) firms that receive bids and sustain a change in control as the basis for emergence from Chapter 11; (2) firms that receive control bids but effect a stand-alone reorganization; and (3) firms that reorganize without having received a control bid. Of the 81 firms that reorganize, 47 (58%) receive control bids. In 37 of these cases (79%) there is a change in control, while in 10 cases (21%) the bid is rejected and the petitioning firm conducts a stand-alone reorganization. These results suggest that when a firm in bankruptcy becomes the target of a bid, there is a high probability that a change in control will occur. In column (1) of Table 6, for the subsample of petitioning firms that trade at the control bid announcement, the three-day average excess returns are 33.11% (t-statistic of 13.21) for 17 firms that sustain a subsequent change in control, and 11.21% (t-statistic of 3.53) for 8 firms with a failed bid. The difference in means calculated t-value for these two sets of firms is 2.33 (p-value = 0.03). This result suggests that control bid announcements are value-enhancing, but less value is created by bids that fail relative to bids that induce a control change. Post-event excess returns, reported in columns (2) through (4), indicate that firms that sustain a change in control retain the gains in wealth generated at announcements of control bids, while gains for failed bids dissipate within three months of the bid announcement. Thus, the maintenance of the gains in target firm value from bid announcements is contingent on a change in control coming to fruition.

V. CONCLUSIONS

We examine valuation effects for creditors holding private secured and unsecured claims against petitioning firms at Chapter 11 filing announcements, and evaluate these effects by type of bankruptcy resolution and by whether there is a subsequent control bid during bankruptcy proceedings. We also assess these effects on equityholders of petitioning firms. Returns to unsecured private creditors, and equityholders of petitioning firms, are significantly negative at Chapter 11 filing announcements, but are less adverse for events in which petitioning firms ultimately reorganize rather than liquidate. Returns to private creditors are also less adverse for events in which the petitioning firm becomes a target of a control bid during bankruptcy proceedings. Losses are consistently more severe for trade creditors than for banks holding unsecured claims. Secured

Table 6. Average Excess Returns at Announcements of Control Bids for Firms in Chapter 11 Proceedings.[1]

	Three-day Event Period	Event Periods		
	−2 to 0 (1)	−2 to 20 (2)	−2 to 40 (3)	−2 to 60 (4)
Control Change N = 17	**33.11%**	**20.32%**	**25.12%**	**31.54%**
	$(13.21)^3$	$(3.16)^3$	$(2.76)^3$	$(2.83)^3$
	[0.88]	[0.76]	[0.82]	[0.82]
No Control Change N = 8	**11.21%**	**7.95%**	**4.14%**	**−5.78%**
	$(3.53)^3$	(0.97)	(0.38)	(−0.41)
	[0.88]	[0.75]	[0.62]	[0.62]
Total Sample N = 25	**26.10%**	**16.23%**	**18.56%**	**19.94%**
	$(13.64)^3$	$(3.32)^3$	$(2.68)^3$	$(2.35)^2$
	[0.88]	[0.76]	[0.76]	[0.76]

Notes: [1] Three-day average prediction errors at announcements of control bids for firms in Chapter 11 proceedings that subsequently reorganize, and cumulative daily post-control bid prediction errors, where prediction errors are calculated using a market model over the pre-event period −240 to −121 prior to the bid, and day 0 is the date of the control bid announcement; t-statistics are in parentheses, the proportion of returns positive is in brackets, and N is the sample size.
[2] Statistical significance at the 5% level.
[3] Statistical significance at the 1% level.

bank creditors earn zero excess returns regardless of bankruptcy outcome or subsequent control bid status of the petitioning firm. Our results are consistent with Diamond's (1993) contention that banks perform a key role in dealing with poorly performing firms and that bank loans, although short term, are effectively senior debt. We also find that the market for corporate control is a significant element of the contemporary bankruptcy process, even though Chapter 11 imposes an environment with considerable impediments to bidders. Bids for control of bankrupt firms are common and generate significant gains to both bidders and target firm equityholders. Most of these bids lead to a change in control for the bankrupt firm. Our evidence suggests that amendments to the Bankruptcy Code that would enhance bidder activities in the market for corporate control and facilitate claims trading would have beneficial effects on the value of petitioning firms and reduce the negative wealth effects of Chapter 11 filings on unsecured private creditors.

REFERENCES

Aghion, P., & Bolton, P. (1992). An incomplete contracts approach to financial contracting. *Review of Economic Studies*, *59*, 473–494.

Aharony, J., Jones, C., & Swary, I. (1980). An analysis of risk and return characteristics of corporate bankruptcy using capital market data. *Journal of Finance*, *35*, 1001–1016.

Baird, D. (1986). The uneasy case for bankruptcy reorganizations. *Journal of Legal Studies*, *15*, 127–147.

Baird, D., & Jackson, T. (1984). Corporate reorganizations and the treatment of diverse ownership interests. *University of Chicago Law Review*, *51*, 97–130.

Betker, B. (1995). Management's incentives, equity's bargaining power, and deviations from absolute priority in Chapter 11 bankruptcies. *Journal of Business*, *68*, 161–183.

Bradley, M., Desai, A., & Kim, E. (1988). Synergistic gains from corporate acquisitions and their division between the stockholders of target and acquiring firms. *Journal of Financial Economics*, *21*, 3–40.

Bradley, M., & Rosenzweig, M. (1992). The untenable case for Chapter 11. *Yale Law Journal*, *101*, 1043–1095.

Byrd, J., & Hickman, K. (1992). Do outside directors monitor managers? Evidence from tender offer bids. *Journal of Financial Economics*, *32*, 195–222.

Clark, K., & Ofek, E. (1994). Mergers as a means of restructuring distressed firms: An empirical investigation. *Journal of Financial and Quantitative Analysis*, *29*, 541–565.

Clark, T., & Weinstein, M. (1983). The behavior of the common stock of bankrupt firms. *Journal of Finance*, *38*, 489–504.

Cornett, M., & De, S. (1991). Medium of payment in corporate acquisitions: Evidence from interstate mergers. *Journal of Money, Credit and Banking*, *23*, 767–776.

Cornett, M., & Tehranian, H., (1992). Changes in corporate performance associated with bank acquisitions. *Journal of Financial Economics*, *31*, 211–234.

Dewatripont, M., & Tirole, J. (1994). A theory of debt and equity: Diversity of securities and manager-shareholder congruence. *Quarterly Journal of Economics*, *109*, 1027–1054.

Diamond, D. (1984). Financial intermediation and delegated monitoring. *Review of Economic Studies*, *52*, 393–414.

Diamond, D. (1993). Seniority and maturity of debt contracts. *Journal of Financial Economics*, *33*, 341–368.

Dodd, P. (1983). The market for corporate control: A review of the evidence. *Midland Corporate Finance Journal*, *1*, 6–20.

Fortgang, C., & Mayer, T. (1990). Trading claims and taking control of corporations in Chapter 11. *Cardozo Law Review*, *12*, 1–115.

French, K., & McCormick, R. (1984). Sealed bids, sunk costs, and the process of competition. *Journal of Business*, *57*, 417–471.

Giliberto, S., & Varaiya, N. (1989). The winner's curse and bidder competition in acquisitions: Evidence from failed bank auctions. *Journal of Finance*, *44*, 59–76.

Hotchkiss, E. (1995). Post-bankruptcy performance and management turnover. *Journal of Finance*, *50*, 3–21.

Hotchkiss, E., & Mooradian, R. (1997). Vulture investors and the market for control of distressed firms. *Journal of Financial Economics*, *43*, 401–432.

Jackson, T., & Scott, R. (1989). On the nature of bankruptcy: An essay on bankruptcy sharing and the creditors' bargain. *Virginia Law Review*, *75*, 155–204.

James, C., & Wier, P. (1987). An analysis of FDIC failed bank auctions. *Journal of Monetary Economics*, *20*, 141–153.

Jensen, M. (1991). Corporate control and the politics of finance. *Journal of Applied Corporate Finance*, *4*, 13–33.

Lang, L., & Stulz, R. (1992). Intra-industry competition and contagion effects of bankruptcy announcements: An empirical analysis. *Journal of Financial Economics*, *32*, 45–60.

Lopucki, L., & Whitford, P. (1990). Bargaining over equity's share in the bankruptcy reorganizations of large, publicly held companies. *University of Pennsylvania Law Review*, *139*, 125–196.

Morse, D., & Shaw, W. (1988). Investing in bankrupt firms. *Journal of Finance*, *43*, 1193–1206.

Rajan, R. (1992). Insiders and outsiders: The choice between informed and arm's length debt. *Journal of Finance*, *47*, 1367–1400.

Smith, J. (1987). Trade credit and informational asymmetry. *Journal of Finance*, *42*, 863–872.

Sushka, M., & Bendeck, Y. (1988). Bank acquisitions and stockholders' wealth. *Journal of Banking and Finance*, *11*, 551–562.

THE WEALTH EFFECTS OF BOARD COMPOSITION AND OWNERSHIP STRUCTURE IN INTERNATIONAL ACQUISITIONS

Sridhar Sundaram, Indudeep Chhachhi and
Stuart Rosenstein

ABSTRACT

This paper examines the association between board composition, ownership structure, and the shareholder wealth of bidding U.S. firms in international acquisitions. Foreign acquisitions represent major investment proposals that demand the involvement of the board in the decision-making process. Hence, the international acquisition process provides a good vehicle to examine the efficacy of the board of directors. Consistent with other studies, our results indicate that U.S. bidders experience significant negative wealth effects at the announcement of the acquisition. In addition, we find that wealth effects are significantly and positively related to board size and to share ownership by independent outside directors and inside directors. Sub-samples based on board size reveal that for small boards, share ownership by directors is significantly related to abnormal returns only for insider dominated boards. For large boards, ownership by independent outside directors is significantly related to abnormal returns only for outsider dominated boards.

Advances in Financial Economics, Volume 6, pages 209–221.
Copyright © 2001 by Elsevier Science B.V.
All rights of reproduction in any form reserved.
ISBN: 0-7623-0713-7

I. INTRODUCTION

The separation of ownership and management in corporations results in potential conflicts of interests between shareholders and managers. The board of directors (BOD) is charged with mitigating these conflicts and protecting and promoting shareholders' interests. To effectively discharge its duties, the BOD is assigned "the power to hire, fire, and compensate the top-level managers and to ratify and monitor important decisions" (Fama & Jensen, 1983, p. 311).

Corporate boards generally include some of the firms' senior managers (inside directors) and a number of outside members (outside directors). Inside directors provide valuable firm-specific information, and outside directors bring objectivity and expertise in decision-making to the board. It is outside directors who can ask the most difficult questions in examining managerial proposals. Thus, it can be argued that shareholder interests are best served by a board independent of management, with a majority of outside directors.

Alternatively, discipline may already be imposed on managers by the market for corporate control, and the BOD may be strongly influenced by management through management's selection of board members and through selective disclosure of information. Thus, it is conceivable that the proportion of outside directors has a marginal impact on managerial performance.

Recent studies find that board composition does matter (see Baysinger & Butler, 1985; Brickley & James, 1987; Weisbach, 1988; Rosenstein & Wyatt, 1990; Byrd & Hickman, 1992). Evidence shows that independent outside directors, who have no affiliation with the firm, represent the monitoring component of the board and promote shareholders' interests.

This study extends the empirical literature by examining the association between BOD composition, ownership structure, and the shareholder wealth of bidding firms in foreign acquisitions. Specifically, we analyze a sample of foreign acquisitions by U.S. firms. Foreign acquisitions represent major investment proposals that allow the transfer or expansion of the firms' resources internationally. Kogut (1983) claims "the primary advantage of the multinational firm, as differentiated from a national corporation, lies in the flexibility to transfer resources across borders through a globally maximizing network." Hence, international acquisitions represent important strategic decisions that require the involvement of the board in the decision-making process. Furthermore, Weiss (1991) indicates that reviewing acquisition proposals put forth by management is a duty of the board.

However, managers may pursue acquisitions for reasons other than shareholder wealth maximization (for example see Jensen & Meckling, 1976; Roll, 1986), requiring outside directors to monitor managerial objectives and to protect shareholders' interests. Hence, the international acquisition process provides a good vehicle to examine whether board composition has an impact on shareholder wealth.

The remainder of the paper is organized as follows: Section II discusses prior research. Testable hypotheses are developed in Section III; the sample and empirical methods are described in Section IV. Results are reported in Section V, and the paper is summarized in Section VI.

II. PRIOR RESEARCH

Board Composition and Shareholder Wealth Effects

Most studies that have examined the role of BODs in the corporate governance process have examined the impact of board composition on firm performance. For example, Baysinger and Butler (1985) report that firms with outsider-dominated boards perform better than firms with insider-dominated boards. Weisbach (1988) finds that boards dominated by outsiders are more likely to force the resignation of poorly performing executives. With respect to corporate combinations, Brickley and James (1987) examine bank mergers and find that the number of outside directors are significantly lower on boards of banks located in states with restrictions in bank acquisitions. This suggests that boards play an important role in evaluating merger proposals. Byrd and Hickman (1992) find that when a firm first announces a tender offer, the share price response of the bidding firm is significantly higher when the board has a majority of independent outside directors. This evidence is consistent with independent boards more carefully choosing and bidding on acquisition targets. Lee, Rosenstein, Rangan and Davidson (1992) find that independent boards monitor managers on behalf of shareholders when examining management buyouts. They find that when management bids for the entire firm, the increase in shareholder gains is significantly higher if the board is dominated by independent outside directors. Brickley, Coles and Terry (1992) examine the shareholder wealth effects of the adoption of poison pill anti-takeover plans. They find that firms with independent boards earn significant positive announcement date returns while other firms experienced significant negative returns. Cotter, Shivdasani and Zenner (1993) examine the role of outside directors in monitoring a tender offer target's managers during the tender offer process. They find that targets with insider-dominated boards are more likely to be resisted, are less likely to be successful, and lead to lower shareholder wealth effects. The above evidence supports the notion that shareholders are better served by boards dominated by independent outside directors.

Mergers and Shareholder Wealth

The empirical evidence on domestic acquisitions is comprehensively reviewed by Jensen and Ruback (1983) and Jarrell, Brickley and Netter (1988). They find that target firm shareholders of successful tender offers and mergers earn significantly positive abnormal returns around the announcement date. Their results also indicate that bidders in successful tender offers enjoy positive returns, but returns are zero for unsuccessful mergers.

In recent years, several studies have examined the shareholder wealth effects of international mergers. Doukas and Travlos (1988) report evidence on bidders' wealth in international acquisitions by U.S. firms. They find that multinational corporations (MNCs) with no previous presence in the target firms' country earned significant positive abnormal returns on the announcement of acquisitions, while other international acquisitions experienced insignificant returns. Conn and Connell (1990) examine a sample of US-UK mergers and report that bidders experienced either positive or zero returns and targets earned significant positive returns. Mathur, Rangan, Chhachhi and Sundaram (1994) examine the wealth effects of foreign bidders in the U.S. and find that shareholders experience significant negative returns. This may be partially explained by the finding that foreign acquirers pay higher premiums than do U.S. firms (Harris & Ravenscraft, 1991).

III. TESTABLE HYPOTHESIS

The literature indicates that mergers significantly enhance the wealth of target firm shareholders. At the same time bidders, on average, experience very small gains or losses. This disproportionate distribution of wealth is also observed in international mergers. Most studies examining the wealth effects of bidders find managerial behavior as a possible explanation for the unfavorable impact of mergers on bidder's wealth (see Jensen & Meckling, 1976; Donaldson & Lorsch, 1983; Roll, 1986). The evidence discussed in the previous section indicates that independent outside directors serve as the most effective monitoring component in a corporate board. Despite the significant growth in international acquisitions by U.S. firms, the effects of board composition on wealth effects of international bidders is unexplored. International acquisitions are important strategic decisions that necessitate the involvement of the board in the decision making process. In this study we extend the empirical literature by examining the BOD effectiveness in monitoring managers of U.S. firms engaging in international acquisitions. Specifically, we test the hypotheses that abnormal returns will be higher for bidders whose boards are dominated by independent outside directors, and for bidders with higher stock ownership by directors.

IV. SAMPLE AND METHODOLOGY

Sample and Data

The sample for this study includes U.S. firms engaging in foreign acquisitions during the eleven year period 1986-96. The initial sample of 221 acquisitions was obtained from the Securities Data Company (SDC). A year-by-year breakdown of the sample is given in Table 1. The precise announcement date is obtained from the *Wall Street Journal Index*. Daily returns data for the sample firms were obtained from the Center for Research and Security Prices tapes. Sample firms were restricted to those listed on the NYSE or AMEX. Data on the board of directors were collected from proxy statements for the regular annual meeting preceding each announcement. The final sample, limited mainly by proxy statement availability, is 194 acquisitions by U.S. firms.

Table 1. Number of Transactions by Year.

Year	Number of Observations
1986*	1
1987	6
1988	9
1989	22
1990	15
1991	11
1992	18
1993	22
1994	25
1995	40
1996**	25
Total	194

* Mergers announced from October 1, 1986 to December 31, 1986.
** Mergers announced from January 1, 1996 to October 28, 1996.

Board Composition

Following previous studies (e.g. Byrd & Hickman, 1992; Rosenstein & Wyatt, 1990), we classify the board of directors into insiders, affiliated outsiders, and independent outsiders, with the classifications closely following Lee, Rosenstein, Rangan, and Davidson (1992). Board composition statistics are reported in Table 2. Panel A reports a mean board size of 11.98 for the sample firms. The average board is comprised of 3.48 insiders, 1.44 affiliated outsiders and 7.06 independent outsiders. Panel B classifies the independent outsiders into three types: financial outsiders, corporate outsiders and neutral outsiders. The table indicates that the average board has a majority of independent directors and among them the majority are corporate outsiders (2.58) and neutral outsiders (3.69).

In addition to board composition, prior studies find stock ownership by directors to be important in alleviating agency conflicts. The statistics on shareholdings by directors is reported in Table 3. For our sample of firms the average inside ownership is about 3.5%, compared with 0.95% ownership by independent outside directors. While independent outside directors hold the majority of seats, inside directors have much higher stock ownership in firms, which should help align the interests of managers and shareholders. Further analysis reveals that average inside ownership is 7.6% for firms where independent outsiders are a minority, compared to just 1.67% for firms where

Table 2. Board Composition.

A. Inside, Affiliated Outside, and Independent Outsiders

	Mean	Median	Minimum	Maximum
Board Size	11.98	12	5	20
Insiders	3.48	3	1	12
Affiliated Outsiders	1.44	1	0	9
Independent Outsiders	7.06	7	1	15

B. Classification of Independent Outsiders

	Mean	Median	Minimum	Maximum
Financial Outsiders	0.79	1	0	5
Corporate Outsiders	2.58	2	0	7
Neutral Outsiders	3.69	4	0	11

Table 3. Percentage of Common Shares Outstanding Held by Inside, Affiliated, and Independent Directors.

	Mean	Minimum	Maximum
Owned by Inside Directors	3.500	0.000	79.974
Owned by Affiliated Outsiders	0.646	0.000	34.856
Owned by Independent Outsiders	0.949	0.000	24.638
Owned by all Officers and Directors	6.990	0.020	81.111

independent outsiders are a majority (not reported in a table). This suggests that the higher insider ownership in insider-dominated boards compensates for the smaller percentage of outside directors, serving to reduce agency problems.[1]

Empirical Methods

We employ standard event study methodology to determine the abnormal returns surrounding the announcement of the foreign acquisition by U.S. firms. The first announcement of the foreign acquisition in the *WSJ* is used as day '0' for event tests. A 100 day trading period from −120 to −21 relative to the announcement is used to estimate the market model parameters. Cumulative abnormal returns (CAR) are computed for various intervals in the period −20 to +20.

We then use cross-sectional regressions to test hypotheses regarding the effects of board composition and ownership structure on the CAR. The variables used to capture board composition and share ownership are defined as follows:

SIZE = size of the board,

INDBOARD = 1 if the proportion of independent outside directors exceeds 50 % of total directors; else 0,

SHARESIND = the proportion of total shares outstanding held by independent outsiders,

SHARESINS = the proportion of total shares outstanding held by insiders,

SHARESAO = the proportion of total shares outstanding held by affiliated outsiders.

Table 4. Cumulative Abnormal Returns (CARs) with Tests of Significance for U.S. Bidders in Case of Cross-Border Acquisitions.

Test Windows	CARs (%)	t-Statistic
(−20 to +20)	0.031	0.115
(−5 to +5)	0.069	0.349
(−1 to +1)	0.091	0.163
(−1 to 0)	−0.166	−1.699**
(0)	−0.193	−1.629**

** Significant at the 5% level.

Table 5. Cross-sectional Regressions for the Entire Sample.

Dependent Variable: $CAR_{-1\ to\ +1}$
(t-Statistics in Parentheses)

	1	2
Intercept	0.0040	0.0084
	(0.929)	(2.117)**
SIZE	−0.0005	−0.0007
	(−1.731)*	(−2.438)**
INDBOARD	0.0028	0.0012
	(1.238)	(0.537)
SHARESIND	0.0549	0.0582
	(1.680)*	(1.768)*
SHARESINS	0.0229	—
	(1.887)*	—
SHARESAO	—	−0.0403
	—	(−1.220)
F	3.181	2.632
p-value	0.0153	0.0365
Adjusted R^2	0.0533	0.0404

** Significant at the 5% level.
* Significant at the 10% level.

We run regressions using three day abnormal returns surrounding the announcement period [CAR (–1,+1)] as the dependent variable.

V. EMPIRICAL RESULTS

The cumulative abnormal returns (CAR) for various intervals surrounding the announcement are reported in Table 4. Consistent with prior literature, significant negative returns are observed for the U.S. bidders. Significant negative CAR of –0.193% and –0.166% are reported for test periods (0) and (–1 to 0) respectively. Similar results are found by Mathur, Rangan, Chhachhi and Sundaram (1993) in the case of foreign bidders acquiring firms in the U.S.

The results from the regression analysis for the entire sample are reported in Table 5. The variable SIZE reveals a significant negative relation between board size and acquisition announcement returns in both regressions 1 and 2. This suggests that for the sample of U.S. bidders, firms with large boards of directors experienced higher abnormal announcement returns than firms with smaller boards. However, the dummy variable INDBOARD, which assumes the value of 1 when the BOD is numerically dominated by independent directors, is not significantly related to the announcement returns.

The coefficients of variables measuring the stock ownership of both inside and outside directors are positive and significant. These results reveal that U.S. bidders experienced negative abnormal returns following the announcement of the foreign acquisitions, but that abnormal returns are significantly related to stock ownership by directors. Large share ownership by both independent outside directors and inside directors are associated with higher abnormal returns. This evidence supports the contention that substantial stock ownership by inside directors mitigates agency problems between managers and shareholders. The results also show an insignificant negative relation between both stock ownership by affiliated outside directors and announcement returns.

As noted previously, in our sample, share ownership by inside directors and independent outside directors varies greatly, depending on whether the board is numerically dominated by independent directors. Given this data, along with the importance of the size of the board, the above results are reexamined by sub-sampling firms based on both the percentage of independent outside directors and board size.

Table 6 reports the cross-sectional results for smaller boards. Smaller boards are defined as boards with eleven or fewer members (fewer than the mean of 11.98). Regressions 1 and 2 of Table 6 report the cross-sectional results for smaller boards dominated by independent outside directors. The results reveal that firms with outsider dominated boards have no significant relation between

Table 6. Cross-sectional Regressions for Firms Where The Size of the Board is ≤ 11.

Dependent Variable: $CAR_{-1 \text{ to } +1}$
(t-statistic in parentheses)

	Percentage of Independent Outsiders > 50%		Percentage of Independent Outsiders ≤ 50%	
	1	2	3	4
Intercept	0.0036	0.0036	−0.0053	0.0007
	(1.318)	(1.400)	(−1.595)	(0.205)
SHARESIND	−0.0835	−0.1721	0.2485	0.2226
	(−0.744)	(−1.368)	(2.899)***	(2.355)***
SHARESINS	0.01159	—	0.333	—
	(0.278)	—	(2.504)***	—
SHARESAO	—	0.2324	—	0.0544
	—	(1.439)	—	(0.956)
F	0.32	01.327	6.976	4.192
p-value	0.7280	0.2748	0.0053	0.0310
Adjusted R^2	−0.0280	0.0129	0.3627	0.2331

*** Significant at the 1% level.
** Significant at the 5% level.
* Significant at the 10% level.

share ownership by directors and announcement returns. This suggests that firms with small boards and with majority independent outside directors on their boards rely primarily on these directors to serve as monitoring mechanism and to mitigate any conflict arising between the managers and shareholders. In contrast, regressions 3 and 4 of Table 6 reveal that small, insider-dominated boards show a significantly positive relationship between stock ownership by independent outside directors and inside directors and announcement returns. These results suggest that stock ownership by directors serves as an important mechanism in overcoming agency issues for smaller boards dominated by inside directors.

Table 7 reports results for boards with twelve or more members. Regressions 1 and 2 provide results for outsider-dominated and insider-dominated boards respectively. These regressions show that for large boards dominated by independent outsiders, stock ownership by these directors are positively and

Table 7. Cross-sectional Regressions for Firms Where The Size of the Board is ≥ 12.

Dependent Variable: $CAR_{-1 \text{ to } +1}$
(t-statistic in parentheses)

	Percentage of Outside Directors > 50%		Percentage of Outside Directors ≤ 50%	
	1	2	3	4
Intercept	-0.0007 (-0.518)	-0.0006 (-0.501)	-0.0007 (-0.227)	-0.0005 (-0.215)
SHARESIND	0.0534 (2.020)**	0.0522 (1.973)*	-0.5400 (-1.082)	-1.2151 (-0.997)
SHARESINS	0.0199 (0.370)	— —	0.0155 (-0.221)	— —
SHARESAO	— —	0.0610 (0.630)	— —	0.2269 (0.602)
F	2.129	2.269	0.621	0.786
p-value	0.1288	0.1132	0.5461	0.4675
Adjusted R^2	0.0388	0.0434	-0.313	-0.0174

*** Significant at the 1% level.
** Significant at the 5% level.
* Significant at the 10% level.

significantly related to announcement returns. In contrast, regressions 3 and 4 indicate that for large insider dominated boards, there is no significant relation between stock ownership and announcement returns.

VI. SUMMARY AND CONCLUSIONS

The board of directors is charged with the responsibility of monitoring managerial behavior and promoting shareholders' interests. Prior studies examining board of directors effectiveness in monitoring managers report that the independent outside directors are the primary monitoring component in a corporate board. In this study, we examine whether board composition and ownership structure affects bidding shareholder wealth. Specifically, we test the hypothesis that U.S. bidders with larger proportions of independent outside directors on their boards and higher stock ownership by directors should experience higher returns upon the announcement of a foreign acquisition. The results indicate that although U.S. bidders experience negative wealth effects following the announcement, abnormal returns surrounding the acquisition announcement are significantly higher for firms with a majority of independent outsiders on the board of directors. The results also indicate that for smaller boards that are numerically dominated by inside directors, a significant positive relation exists between stock ownership of inside directors and independent outside directors. For firms that are numerically dominated by independent outsiders, a significant relation is observed between stock ownership and the announcement returns only for large firms. This supports the claim that in small boards dominated by inside directors, stock ownership is used as an important mechanism to mitigate conflicts of interest between management and shareholders. In contrast, for firms with large boards, share ownership by independent outside directors is important only when the board is dominated by independent outside directors. For insider dominated large boards, no significant relation between stock ownership and announcement returns is found.

NOTES

1. We also collected data on director tenure, CEO duality, existence of a staggered board, number of board meetings per year, the presence of other block owners, and management compensation. These variables were eliminated from the analysis because preliminary analysis indicated that none of these variables were significantly related to abnormal returns.

REFERENCES

Baysinger, B., & Butler, H. (1985). Corporate governance and the board of directors: Performance effects of changes in board composition. *Journal of Law, Economics, and Organization, 1*, 101–124.

Brickley, J., & James, C. (1987). The takeover market, corporate board composition, and ownership structure: The case of banking. *Journal of Law and Economics, 30*, 161–181.

Brickley, J., Coles, J., & Terry, R. (1994). The board of directors and the enactment of poison pills. *Journal of Financial Economics, 35*, 371–390.

Byrd, J., & Hickman, K. (1992). Do outside directors monitor managers? *Journal of Financial Economics, 32*, 195–221.

Conn, R., & Connell, F. (1990). International mergers: Returns to U.S. and British firms. *Journal of Business Finance and Accounting, 17*, 689–710.

Cotter, J., Shivdasani, A., & Zenner, M. (1993). The effect of board composition and incentives on the tender offer process. University of Iowa, working paper.

Donaldson, G., & Lorsch , J. (1983). *Decision making at the top: The shaping of strategic direction*. New York: Basic Books.

Doukas, J., & Travlos, N. (1988). The effect of corporate multinationalism on shareholders' wealth: Evidence from international acquisitions. *Journal of Finance, 43*, 1161–1175.

Fama, E., & Jensen, M. (1983). The separation of ownership and control. *Journal of Law and Economics*, *26*, 301–325.

Harris, R., & Ravenscraft, D. (1991). The role of acquisitions in foreign direct investment: Evidence from the U.S. stock market. *Journal of Finance*, *46*, 825–844.

Jarrell, G., Brickley, J., & Netter, J. (1988). The market for corporate control: The empirical evidence since 1980. *Journal of Economic Perspectives*, *2*, 49–68.

Jensen, M., & Meckling, W. (1976). Theory of the firm: Managerial behavior, agency costs, and ownership structure. *Journal of Financial Economics*, *3*, 305–360.

Jensen, M., & Ruback, R. (1983). The market for corporate control. *Journal of Financial Economics*, *11*, 5–50.

Kogut, B. (1983). Foreign direct investment as a sequential process. In: C. Kindleberger and D. Andretsch (Eds), *The Multinational Corporation in the 1980s* (pp. 38–56). Cambridge, MA: The MIT Press.

Lee, C., Rosenstein, S., Rangan, N., & Davidson, W. (1992). Board composition and shareholder wealth: The case of management buyouts. *Financial Management*, *21*, 58–72.

Mathur, I., Rangan, N., Chhachhi, I., & Sundaram, S. (1994). International acquisitions in the United States: Evidence from returns to foreign bidders. *Managerial and Decision Economics*, *15*, 107–118.

Roll, R. (1986). The hubris hypothesis of corporate takeovers. *Journal of Business*, *59*, 197–216.

Rosenstein, S., & Wyatt, J. (1990). Outside directors, board independence, and shareholder wealth. *Journal of Financial Economics*, *26*, 175–192.

Weisbach, M. (1988). Outside directors and ceo turnover. *Journal of Financial Economics*, *20*, 431–460.

Weiss, E. (1991). The board of directors, management, and corporate takeovers: Opportunities and pitfalls. In: A. Sametz (Ed.), *The Battle for Corporate Control*. Homewood, IL: Business One Irwin.

CONCORDIA UNIVERSITY LIBRARIES
SIR GEORGE WILLIAMS CAMPUS
WEBSTER LIBRARY

DISCARDED
CONCORDIA UNIV. LIBRARY

CONCORDIA UNIVERSITY LIBRARIES
SIR GEORGE WILLIAMS CAMPUS
WEBSTER LIBRARY